Praise for The Complete Guide to Seed & N

This beautifully illustrated book gives us the knowledge a.. inspiration to step off the train of corporate mass production and to explore a world of rich flavors and colorful oils. Cohen encourages readers to embrace a deeper and healthier, slow approach to living.

—Anna Mulè, executive director, Slow Food USA

Bevin Cohen takes us on an incredible journey with his easy, authoritative, and nurturing style. This book is a must for anyone interested in food as well as folks interested in self-reliance. Homesteaders, specialty crop farmers, gardeners, chefs, seed-to-table enthusiasts, and curious humans, would all be enriched by a copy of this book held close.

—Hank Will, editor at large, *Mother Earth News* magazine

A must-have for anyone who wants to be self-reliant with their food production, as well as foodies who just love knowing how food is produced.

—Deborah Niemann, author, *Homegrown and Handmade*

Bevin walks you through homemade oil production from seed to press to pantry in this delightfully informative and accessible book.

—Elizabeth Hoover, Associate Professor, Environmental Science, Policy, and Management, UC Berkeley

I am standing in the front row cheering for this amazing book. Cohen understands the facade of modern agriculture, which is really a crisis of agriculture not yet visible in the grocery. This is a deeply important book, a toolkit, a plea, a template. Add it to your library of books that will save us.

—Janisse Ray, author, *The Seed Underground: A Growing Revolution to Save Food*

Beautifully illustrated and photographed, this guide is sure to introduce readers to new concepts and whet their appetite for growing plants with a whole new purpose. It's a must read for gardeners everywhere!

—Brie Arthur, author, *The Foodscape Revolution* and *Gardening with Grains*

Cohen strives to wrest the production of plant-based oils from industrial processors with this treasury of practical observation and honest, clear-eyed recommendations. There is much learning here—historical, chemical, and medicinal. This is the comprehensive handbook you've been wishing for.
—Dr. David S. Shields, agricultural and culinary historian

Invaluable to anyone wishing to process their own seed and nut oils. It is definitely the most complete guide that I have seen! Bevin's detailed explanations and precise instructions are easy enough for anyone to follow.
—Jere Gettle, founder, Baker Creek Heirloom Seed Company

Homesteaders and producers alike recognize that freshly grown and processed foods, by our own means, have the best flavor and nutritional value. Bevin Cohen unlocks the mysteries of nut oils, one of the last frontiers for DIYers, and empowers the reader with his passion and knowledge to tackle fresh pressed oils.
—Julia Shanks, co-author, *The Farmers Market Cookbook*, author, *The Farmer's Office*

A nourishing capsule of informative beauty that will inspire readers to dig deeper into the alchemical world of plants to discover, and hopefully revive, traditional artisan skills.
—Chrissie Orr, co-founder, SeedBroadcast

There is a lot to learn within these pages. Bevin has crafted an easy-to-use guide that is sure to become a frequently referenced book. Discover how simple it can be to press your own oils from the seeds in your garden and reclaim this traditional artisan skill.
—Sonya Harris, Founder and CEO, The Bullock Garden Project, Inc.

Intentionally holistic, practical and accessible, the abundant information in this book obviously comes from Cohen's depth of personal experience. With clear, easy to follow explanations and beautiful artwork and photography, this guide is sure to be a useful reference for anyone interested in learning the science and art of pressing seed and nut oils at home.
—Melissa DeSa, Working Food

THE COMPLETE GUIDE TO
SEED & NUT OILS

GROWING, FORAGING, AND PRESSING

BEVIN COHEN

new society
PUBLISHERS

Cover design by Diane McIntosh.

Drawings: Alicia Mann.

Photos credits: Heather Cohen (unless otherwise noted). Page 1: alex9500;
p. 54: bob; p. 69: smallredgirl; p. 104: dtatiana; p. 108: dule964, Igor Syrbu, geografika;
p. 109: J and S Photography, Mara Zemgaliete; p. 110: vainillaychile, Dzha, kovaleva_ka;
p. 111: vladimir18, kitsananan Kuna; p. 112: Amy Lv, Iurii Kachkovskyi, seqoya;
p. 113: Voravuth, Evgeny, Virtexie; p. 114: Chris Redan, Yutthasart; p. 115: Jogerken,
wasanajai; p. 116: Elena Schweitzer, dolphfyn; p. 117: Elena Schweitzer, Pineapple
studio; p. 118: Agave Studio, sommai, ksena32, Dmitriy / Adobe Stock.

Printed in Canada. Second printing August, 2023.

Inquiries regarding requests to reprint all or part of *The Complete Guide
to Seed & Nut Oils* should be addressed to New Society Publishers at the address
below. To order directly from the publishers, please phone 250-247-9737
or order online at www.newsociety.com. please order online at
www.newsociety.com

Any other inquiries can be directed by mail to
New Society Publishers
P.O. Box 189, Gabriola Island, BC V0R 1X0, Canada
(250) 247-9737

LIBRARY AND ARCHIVES CANADA CATALOGUING IN PUBLICATION

Title: The complete guide to seed & nut oils : growing, foraging, and pressing /
Bevin Cohen.

Other titles: Complete guide to seed and nut oils

Names: Cohen, Bevin, 1979– author.

Description: Includes index.

Identifiers: Canadiana (print) 20210339241 | Canadiana (ebook) 20210339276 |
ISBN 9780865719637 (softcover) | ISBN 9781550927566 (PDF) | ISBN 9781771423526 (EPUB)

Subjects: LCSH: Oils and fats, Edible. | LCSH: Seeds. | LCSH: Nuts.
| LCSH: Cooking (Oils and fats)

Classification: LCC TX407.O34 C64 2022 | DDC 641.3/385—dc23

Funded by the Government of Canada	Financé par le gouvernement du Canada

New Society Publishers' mission is to publish books that contribute in fundamental
ways to building an ecologically sustainable and just society, and to do so with the
least possible impact on the environment, in a manner that models this vision.

Contents

To Heather, Elijah and Anakin:
thank you for always reminding me to grow.

Preface

We live in a land of abundance and plenty. Examples of this substantial bounty can be found in various locales, but perhaps no other place highlights our good fortune quite like the sprawling American supermarket. With dozens of aisles, stacked floor to ceiling, loaded with fresh fruits and vegetables, breads and beans, processed and packaged foods, herbs and spices from faraway lands, household products, anything and everything that we may hope to purchase and enjoy, the supermarket has it all. If we approach the grocery store with the eyes of a consumer, we are thrilled by its diversity of offerings. In fact, many consumers have come to expect this abundance, demand it, and are quick to complain to the store's management if the selection available isn't as diverse and lavish as expected.

If we view the market with the eyes of a producer—whether farmer, homesteader, community activist, crafter or artist—we see these aisles of grocery-laden shelves in a very different light. We see the hours of labor spent cultivating, harvesting and processing the crops. We see the daily struggles of the farmer, battling the elements to ensure a fruitful season. We see the processor, crafting raw ingredients into viable commercial products. We see, and are acutely aware of, the sheer amount of resources that are spent, exploited and far too often wasted in a desperate attempt to keep these shelves filled with a myriad of choices, in what too often appears to be an overly aggressive assault on our senses. As if, perhaps, by loading these shelves with dozens of options, countless brands of packages stacked up in neatly organized rows, the store itself is trying to convince us of our wealth. How could we possibly feel insecure or unsure of our standing in the world when we have 24 different brands of breakfast cereal to choose from? But the producer knows better.

Through the eyes of the producer, we see that what appears to be 24 distinct selections to choose from is merely an illusion. Although the choices may have unique packaging with different colors, slogans and mascots, the producer knows that a majority of these products are just varying ratios of corn and sugar. And in most of these cases, the sugar is likely just high-fructose corn syrup.

While the illusion of choice may appeal to the wide-eyed, eager consumer, those of us with a production-centered state of mind find ourselves quickly disenchanted by this overreaching, almost ridiculous, façade. Farmers, homesteaders, community activists, crafters and artists alike are driven by a philosophy of creation, a true do-it-yourself mindset. We will not, we cannot, just sit back contently and allow others to produce the world's goods for still others to consume, because a system based on unlimited consumption, with limited production, simply cannot sustain itself. And certainly not with the quality that we so desire. When as individuals, or cooperative communities, we commit to a focus on production, creating for ourselves what we can with the resources available to us, this is when we begin to restore balance to a system that has become stretched so thin it can hardly be expected to support us for much longer.

The realities of self-reliance, to any degree, can certainly seem overwhelming, even to the most energetic and inspired among us, but as with any other endeavor worth pursuing, the realization of our goals is well worth the stress, sacrifices and hard work required to reach them. And like any other journey toward self-improvement, the path forward begins with the first step. None of us will see our dreams manifest overnight but only through a continual process of small steps forward, coupled with self-evaluation and continual redirection, always steering ourselves toward our desired destination. Over time, even our destination may change, but each small step forward brings us that much closer to where we wish to be.

My personal journey toward finding a balance between my perceived need to consume and my ability to produce began many years ago. As a practicing herbalist, I was already keenly aware of my consumer-based dependency on goods produced outside of the family homestead. In an attempt to rectify what I saw as an unsustainable imbalance, I committed myself to shifting this perceived need into action. Why purchase what I could create? If I am unable to produce what I believe that I need, do I, in fact, really need it? Or is there an alternative?

To improve upon a system, we must first evaluate its flaws. This doesn't mean that we shouldn't recognize its strengths as well, but this latter task is generally a far quicker and more comfortable one. Identifying our flaws, be they personal, professional or societal, requires honesty and reflection, and this often leads to realizations that are difficult to embrace. But embrace them we must, if we ever hope to shed this skin of consumerism to emerge as independent, truly sustainable producers.

When faced with these difficult, personally existential questions, I was surprised to find answers in the least likely of all places: the grocery store. I suppose that what I discovered there was not so much direct answers to my questions but the impetus for the train of thoughts that led me to where I now find myself.

As a proponent of local food and a catalyst for positive change, I had long endeavored to scale down and simplify my family's personal food system, growing our own whenever possible and purchasing directly from local producers to supplement our needs. Studying local food systems, I realized that the foundation required by these systems in order to meet the true needs of a community is built upon grains and legumes. These staple crops provide the caloric requirements, and therefore the energy, that a society needs in order to properly function.

The community where I live in central Michigan is fortunate enough to be home to a number of artisan bakers, all skillful and dedicated to their craft. Many of these bakers offer unique breads made from ancient grains such as emmer and einkorn wheats. Some of these artisans source their grains locally, dedicated to the finest flavors that can be achieved only through the freshest ingredients, choosing to mill their own flours for each freshly baked batch of breads.

It was with these artisan bakers in mind, and their devotion to producing the highest quality offerings by milling raw ingredients into viable commercial products with their own hands, that I found myself deep in personal reflection, staring out at the abundance of cooking oils available for purchase at my local supermarket. Just as bread itself has an ancient history, with the first flatbreads being produced many thousands of years ago and leavened breads, those made light and fluffy with yeast, becoming common around 300 BCE, vegetable oils have been a part of human cuisine since antiquity. If producing my own staple crops, such and grains and beans as well as fruits and vegetables, was essential in my quest to avoid needless consumption and gain greater independence, then by that same logic, wouldn't home-scale production of seed and nut oils be a justifiable venture?

Nearly a decade has passed since the fateful day when I decided to try my hand at artisan oil production, and the task has proven both profitable and rewarding. We've now extracted oil from the seeds and nuts of more than a dozen plant species at Small House Farm, and we've enjoyed these oils in our kitchen and used them in our herbal apothecary as the base of numerous topical wellness products. Our small-batch, expeller pressed oils have been sold commercially through numerous cooperative groceries and health food stores, as well as via our online platforms, and have even earned national awards for their quality.

It's my objective that this book will serve as an instructional guide to not only equip you with the knowledge required to successfully press your own seed and nut oils, perhaps even growing your own oilseed crops, but to also help you undertake, or continue along, the often challenging journey away from mindless consumption and toward mindful production.

Before we continue, I must revisit my opening analogy, which portrayed the supermarket as a symbol of abundance and plenty. While this certainly holds true, we must acknowledge that this extravagance is a luxury not equally available to all that may wish to partake. Many factors affect this unfortunate situation, including those of socio-economic inequality, and the fact remains that, indeed, some neighborhoods are devoid of anything resembling a grocery store at all.

Perhaps this somber reality further illustrates my point that whenever possible, with whatever resources available, it is vital that each of us, each individual, family and community, reevaluates our perceived needs, takes stock of our abilities, and pushes forward in every reasonable way to grow our own, produce rather than consume, and to once again regain control of the basic components of life. The quality of our livelihood rests in our own hands; let us use our hands to produce a better world.

Oil Extraction:
A Brief History

PLANT-BASED OILS, extracted from seeds, nuts and occasionally fruits, have been a fundamental part of the human diet since as early as 6000 BCE. Archeologists have uncovered evidence of olive oil production in ruins discovered in northern Israel that are believed to be around 8,000 years old, while in North America, archeologists from Indiana University have found evidence of hickory nut oil extraction in the remnants of an ancient kitchen presumed to be over 4,000 years old. Residue believed to be oil extracted from brassica seeds has been identified inside shells discovered within an eighth- to tenth-century church of a Coptic monastery in Egypt, and historic records indicate that brassica oilseed crops had been cultivated as early as 2000 BCE. These oils were vital to the health and well-being of their producers. The dietary fats found in seed and nut oils play a fundamental role in the body, assisting with the proper absorption of vitamins as well as being essential to brain and nerve function.

Aside from the culinary applications, these oils were also used as fuel, most often in lamps, which were the principal source of lighting in ancient times. Before the advent and widespread utilization of kerosene, animal fat and plant-based oils were the predominant choices for lamp fuel. These lamps were much preferred over candles for lighting, due to the extended period of time that such lamps could be burned. One of the most well-known stories involving the use of oil lamps is, of course, that of Hanukkah, known as the Festival of Lights, a Jewish celebration commemorating the rededication of the Second Temple in Jerusalem; and indeed, both the Old and New Testaments of the Bible record multiple instances of the use of olive oil as fuel as well as food.

Olive oil press at Assissi (1676).

As a quick point of clarification, relevant and worth mention before we can continue, throughout this book, when referring to oil, I am never referring to mineral oils, crude oil, petroleum or any of its refined, petrochemical components. The obvious focus of this volume is plant-based seed and nut oils, their production and various uses.

These oils were certainly important to ancient religions, as can be evidenced by numerous traditional ceremonies, many of which are still observed today. The ritual act of anointing with oil even carried into secular life, as an honor bestowed upon monarchs and government officials and in some cultures even considered a mark of hospitality offered to houseguests. This is not to imply that topical application of oils was limited to significant moments of celebration or spiritual observation. In fact, various oils, often scented with herbs or flowers, were employed as a treatment for those fallen ill or even to simply mask unpleasant odors.

In the 8,000 years since the first olive was squeezed to release its flavorful, golden essence, the use of oils in food, medicine and for religious ceremony hasn't undergone many notable changes. In kitchens around the world oils are still employed as a medium for cooking as well as to improve the flavor of various dishes. They are frequently utilized as the base of commercially manufactured, as well as household, preparations of topical wellness and beauty products, including soaps, lotions and medicinal balms. And while not as common as the practice once may have been, modern religious ceremonies still make use of oils in many of their rituals. Various oils have proven useful for industrial applications as well, as insulators, lubricants and, of course, as a source of fuel. While oil lamps are certainly not as widely used as they once were, numerous vegetable oils are converted into biofuels, which are then utilized to fuel vehicles, to generate energy for utilities such as heat and electricity, and also for cooking.

While the uses for seed and nut oils have remained somewhat consistent over time, the methods of production have evolved dramatically since the first olives were pressed along the Mediterranean Sea so many years ago. It's likely that the earliest technique to extract oil was what is known simply as the wet extraction method. For this procedure, the seeds or nuts are first hulled, then heated and crushed. Next, the ground seeds are boiled in enough water to suspend the crushed seed material until the oils are coaxed free and float to the surface. The oils would then be carefully skimmed from the water and heated again, this time in a smaller container, to evaporate any additional moisture. This simple technique requires little equipment, but is quite time consuming and results in relatively low yields.

Another traditional approach to oil extraction, utilized by cultures around the world, although with modifications in mechanics and design, is that of the manual press. This method makes use of pressure for extraction, essentially squeezing the oils free from the seeds, nuts and fruits. Variations of this technique are what we will focus on throughout this book for their efficiency and ease of application for the small-scale producer.

Historians have determined that around 2000 BCE, on the Indian subcontinent, sesame seeds were commonly pressed for their oils using a stone or granite mortar and pestle. Larger versions of this device were constructed, and young bulls were employed to turn the pestle within the mortar, thus crushing the seeds, their oils running out of a small hole at the base of the mortar to be collected. These machines are known as *ghanis*, and modern, motor-powered models are still widely available, although they remain most commonly used throughout India.

A variation of this technology, still relying upon the principle of applying direct pressure to the oilseed, is commonly known as an expeller press. These machines are composed of a rotating turnscrew, housed within a horizontal cylinder, that gradually increases the pressure on the seed or nut when turned. These machines can be manually operated or motorized, and this category of oil press is the type that we will discuss in most depth in later chapters, it being readily accessible, easy to operate and efficient, while providing the highest yields in comparison with other mechanized oil presses.

A traditional ghani-style press used for pressing peanut oil in India.

Alternatively, devices such a ram press can be used for processing soft seeds such as sesame or sunflower. This machine operates with a piston that is manually forced via a lever into a horizontal chamber, crushing the seeds that have fallen into the chamber from a hopper above. While quite efficient with softer seeds, this type of machines is unable to successfully press hard seeds and can easily be damaged by the attempt.

Another piece of equipment, likely already familiar to the homesteader and farmer, is the cage press. This piece of equipment is a vertical press and is most commonly employed to crush and squeeze the juice from apples, grapes, pears and similar fruits. These machines can be manually operated or motorized, are simple to use and provide good yields. The most notable drawback to this style of oil press is that it operates on a batch system. Only a certain amount of seed can be pressed at a time, and the machine must be cleaned out between each batch to maintain quality and efficient yields, unlike the expeller or ram style presses, which offer continuous operation.

Regardless of the style of extraction equipment used, the oils produced by the methods listed above are considered unrefined, and in cases where pressing occurs at temperatures less than 122°F (50°C), these oils would also be considered cold-pressed. Before we go any further, let's delve into the differences between, as well as the pros and

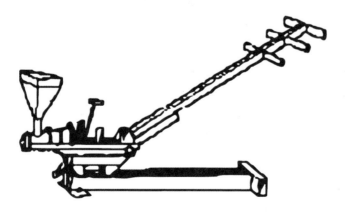

Credit: © Creative Commons www.creativecommons.org

A ram-style oil press.

cons of, refined versus unrefined oils, mechanical extraction versus solvent extraction and why cold-pressed oils are considered superior by health-conscious consumers. We should also discuss the processes of bleaching, deodorizing and hydrogenating that many commercial oils go through, if only to emphasize the necessity for small-scale, environmentally conscious and mindful producers.

The extraction techniques previously discussed are considered mechanical extraction methods; that is, the process of extraction relies upon manual or motorized impact, utilizing pressure to remove the oil from the chosen seed, nut or fruit. The resulting oils are considered unrefined, having endured no further processing. Allowed to rest, any particles that remain in the oil will settle, and the final product can then be decanted or lightly filtered before bottling. These unrefined oils have the most pronounced flavor and color and are considered to be of the highest quality. These oils are typically the most nutritious, but, due to the natural resins and minute particles within are best when used unheated and have a shorter storage life than refined oils. These products can then be more thoroughly strained and filtered to remove additional small particles and improve their shelf life. Gently heating the oil first will make filtering more successful. After this additional

The light color of refined pumpkin seed oil compared to the deep red color of the unrefined oil.

processing, the oil would then be referred to as "naturally refined," and these products tend to have a higher smoke point, making them a better choice for high-heat cooking applications.

Conversely, a great majority of commercial oils are chemically refined. This is a multistep process intended to "purify" the oil by removing any impurities, creating a product with an almost indefinite shelf life. The first step is the degumming process, in which oil is heated and then mixed with either a calculated amount of water or phosphoric acid, depending on the type of oil being refined. The resins present within the oils are thus separated and then removed from the oil, which moves along to the second step in the process: neutralization. This second step, introduction of an alkali, neutralizes free fatty acids and removes the acidity from the oil.

From this point, the oil enters the bleaching process. Here the product is heated to a temperature of 248°–266°F (120°–130°C), and bleaching clay is introduced to remove the color pigments from the oil until it reaches the desired quality oil color for the product. The chemicals are then filtered from the product once this color is reached. The next step is the deodorization process, to remove any odors from the oil. To achieve this, the oil is heated to 392°F (200°C) in a pressurized vacuum as steam is forced through the oil to evaporate any odor-causing substances. After this step some oils, such as sunflower, rapeseed and corn oil, are exposed to the process of dewaxing. This final step is to ensure that the oils remain clean and do not become cloudy when exposed to colder temperatures. To accomplish this, the oils are first heated to 131°F (55°C) and then quickly cooled to 50°–59°F (10°–15°C). Any solids that form during this cooling process are filtered out, and the resulting oil is a clear, odorless liquid.

In addition to the long and arduous refinement procedure, most large-scale commercial oils are processed via chemical extraction, as opposed to the mechanical techniques previously described. This solvent extraction technique most commonly involves the use of petroleum-derived hexane. The oilseed is first crushed or flaked, then blended with the solvent chemical. After the prescribed extraction time, the plant material is removed and the solvent is evaporated from the oil by heating the mixture to a temperature of 300°F (149°C). The resulting oil is then refined as described above. This inexpensive and high-yielding technique has become the most popular and widely used method of oil extraction, particularly in the manufacture of commodity crop oils such as corn, soybean and cottonseed.

An additional process that some commercial oils are exposed to is known as hydrogenation. This is a common practice with oils extracted from soy, sunflower, cottonseed and olive, among others, through which the oil is blended with a catalyst (typically finely ground nickel) and heated in a large-capacity cylindrical pressure reactor to 248°–370°F (120°–188°C). The mixture is continuously stirred while hydrogen gas is pumped through the liquid, creating a chemical reaction. The final, semi-solid, modified product is considered a hydrogenated or partially hydrogenated oil. This technique is meant to improve the flavor, stability and storage qualities of the oil, but these products are typically considered to be nutritionally inferior and are generally avoided by health-conscious consumers who believe the artificially manufactured trans fats found in these products can adversely affect heart health.

Among the various treatments, chemical and otherwise, that oils are subjected to throughout the refinement process, it's the frequent exposure to high temperatures that may potentially be the most destructive aspect for the oils' nutritional values. While manual extraction methods could never produce the extreme heat that's involved in the refinement process, it is possible for the motorized equipment to reach temperatures above the 122°F (50°C) threshold that defines whether or not an oil can be labeled as cold pressed. Oils that, during production, maintain a temperature below this defining threshold will retain maximum flavor, aroma and nutrients and are considered to be of the highest quality. Expeller presses, which squeeze the seeds through a cylindrical cavity with a turnscrew, or ghanis, the motorized mortar and pestle style machines, although not heated, can reach temperatures of up to 140°–210°F (60°–99°C) due to the intense friction and pressure needed to extract the oil. While cold pressing seeds or nuts may produce the highest-quality oils, the yields are measurably less than what can be achieved by utilizing an expeller-style press. In any case, both of these oil products, cold pressed or expeller, are superior in flavor and nutritional value when compared to their solvent-extracted and chemically refined counterparts.

Small-Scale Production:
A How-To Guide

SMALL-SCALE EXTRACTION OF SEED and nut oil is a relatively straightforward process that can easily be integrated into daily life, but the first steps to take are analyzing your needs and defining your goals. Is your plan to simply supplement your family's commercial oil dependency or to replace it entirely? Remember, as with any long-term DIY endeavor, it's important to take things slowly. Understanding your needs, as well as your abilities and available time, will help you to determine the level of commitment, and corresponding amount of labor, that you are willing to devote to this work. When taking stock of your current oil use, make note not only of the volume of oil used but also which types are the most common in your household as well as their purpose. Are these oils being put to use solely in the kitchen, or are some for topical, medicinal or other applications? Develop a comprehensive list that includes all of this information.

Additionally, some thought should to be put into the various species of seeds and nuts that you'll be pressing and from where these crops can be sourced. If your time and space allow, growing your own oilseed crops can be a rewarding and economically sensible option, giving you the greatest control over the quality of your oilseeds. This is certainly the most sustainable, self-reliant arrangement, although growing, harvesting and processing your crops will take an additional commitment of time and labor. As an alternative, sourcing your seeds from local growers still enables you to be somewhat connected to the product while at the same time supporting your local economy. Of course, either of these options is possible only with oilseed crops that can be grown in your region. If your needs require you to press oil from seeds, nuts or fruits that can't be grown in your area, then you'll need to purchase them in bulk from a commercial source. This presents another good opportunity to subjectively analyze your needs: is this particular "foreign" seed or nut oil truly indispensable, or is there an alternative available, a local seed or nut whose oil might make a suitable substitute? In Section Three we'll delve into more specific information on a number of seeds, nuts and fruits that are well suited to quality oil production, including planting, growing and harvesting instructions, as well as any post-harvest processing requirements, details on storage, shelf life and also the uses and health benefits of the various oils. Whether your intention is to pursue small-scale oil extraction as a commercial endeavor, a part-time hobby, or anything in between, this field-to-press style guide will help you determine which oilseeds are most ideal for you, dependent upon your region, your available space, your oil needs and, ultimately, your goals.

Once you've established your initial objectives, the next major step is investing in the oil press. As I mentioned earlier, the expeller-style press, with a horizontal cylinder and rotating turnscrew, will be the machine most thoroughly discussed throughout the book, but we will continue to touch on the pros and cons of other mechanized oil presses, especially once we delve into the individual oilseeds in the next section. Unless you're planning to jump right in to commercial-scale production, a basic, manually operated expeller press will suffice for a start-up operation. Significantly less expensive than their motorized counterparts, manual expeller presses are easy to operate and, as we'll explore later, can also be upgraded if your production needs increase.

A simple machine, the manual expeller press typically consists of only four or five parts: the main body, which mounts onto a table or other sturdy surface and includes the cylinder within which the oilseed is pressed; the turnscrew; the crank used to turn the screw; a hopper to hold the seeds waiting to be fed into the machine; and an adjustment bolt that's used to restrict the gap at the end of the cylinder through which the crushed seed, also known as oilseed cake, is then discharged

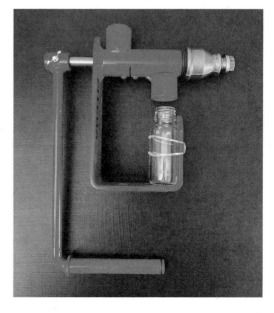

Above: The parts of a typical, hand-turned, expeller style press on display.
Below: A fully constructed, hand-turned, expeller-style oil press.

from the machine. The simplicity of the press allows for easy upkeep and maintenance, which may not always be possible with a motorized model.

Operating the oil press is a straightforward task. Typically, with an expeller-style machine, the first step is to slowly heat the pressing chamber to facilitate the extraction process and increase yields. In the case of a manually operated machine, the heat is generated either with a small candle or kerosene lamp. The height of the flame, and its proximity to the machine, can be adjusted to allow some control over the temperature of the oil, and with observation and readjustment, the temperature can be kept below the 122°F (50°C) benchmark that defines an oil as cold pressed. Motorized presses generally make use of an electric heating element, and many models offer adjustable temperature controls, allowing the operator to maximize the extraction yields for various seeds and nuts.

As can be expected, a motorized oil press will cost more than a manually operated machine, and while less laborious to operate, this larger investment may not be sensible for your situation. The oil needs of most homes simply won't warrant the expense of a motorized press.

A standard hand-turned oil press is upgraded to a motorized model with a pulley and belt.

A manual, hand-turned, expeller press requires only a modest upfront investment, and, over time, if your needs or plans outgrow the production capabilities of the machine, or the stamina of your arm, it's certainly possible to upgrade your manually operated machine to accommodate this. The first approach is to simply mount a motor onto the machine, likely using a rubber belt between the motor shaft and the turnscrew shaft of the oil press. This can be facilitated by the use of a pulley wheel, or sheave, and there are various ways in which the system can be constructed based on the available parts and resources as well as the mechanical aptitude of the engineer.

A second method to upgrade your manual press involves a similar concept and utilizes many of the same components, but instead of an electric motor, it is reliant on pedal power to drive the machine. This technique again uses a belt to operate the machine, but this time the belt will run from the turnscrew shaft of the press to the axle of the moving wheel of a bicycle. Of course, this does involve more physical effort to operate than a motorized model but is still very productive and does not rely on the use of electricity, thus allowing the machine be used anywhere. Regardless of the type of upgrade one might choose, the need for a larger hopper is still an important consideration. Motorized or pedal-powered, these upgraded presses will help you sustain a higher rate of production for an extended period of time and will therefore run through significantly more seed than would be possible using the unaltered, hand-turned machine.

Regardless of whether the press is manually operated or motorized, the process of extracting the oil remains the

An oil press can be easily converted to "pedal-power" using a sprocket and bicycle chain.

A vibration-free motor belt chain is a useful option when upgrading a manual oil press to a motorized model.

same. Some of the seeds, nuts or fruits may require treatment in preparation of pressing, be it shelling, heating or drying, and we will cover the specific needs of each oilseed crop in Section Three. Once the seeds are processed, it's time to ready the press for work. As mentioned earlier, many presses utilize heat to facilitate extraction. This could be as simple as flipping a switch to engage the electric heating element or lighting a small lamp or candle and placing it under the pressing chamber of the machine. If you're using a batch press-style machine, such as a cage press or ghani, you'll skip this step. These other presses rely solely on pressure to create extraction. In the case of a ghani, a machine that functions somewhat like a mortar and pestle, the friction

created by its grinding action will sometimes generate a small amount of heat, but this is negligible in comparison to the heat that's typically applied to the chamber of an expeller-style machine by an electric heating element or kerosene lamp.

When the machine reaches the proper temperature for pressing, load the oilseeds into the hopper and engage the turnscrew. Within moments, oil should begin flowing from the machine, at a rate dependent on the particular seed being pressed, while the remaining organic material, known as seedcake, oil cake or press cake, will be expelled from the end of the machine. This seedcake should not be considered waste; on the contrary, this byproduct of oil production is actually quite useful. Depending on the seed or nut being pressed, these remnants can be used in the kitchen, as livestock feed or for various other applications. The potential uses for these press cakes will also be explored in more depth in the following section.

As the machine runs, the hopper will need to be refilled from time to time. Oil will be collected in a reservoir as it drains from the output slit, and the seedcake should be caught in a bucket or other container as it is expelled from the press.

Freshly pressed hempseed oil and the seedcake resulting from the extraction process.

You can run the machine for as long as you desire, until you run out of seeds to press or whenever you need a break, although many motorized presses, especially the smaller models, have limitations on how long the machine should be run to avoid overheating. Consult your owner's manual for specific requirements and restrictions for your equipment.

Once pressing is completed, the oil you've collected will be visibly turbid and need time to settle. Cover the container, and set it in a cool, dark location somewhere out of the way to rest. Be sure to label the oil with the name of the seed or nut as well as the date of pressing. Within 24 hours, the oil will have visibly clarified, and the debris, mostly fine particles of seed or nut, will have accumulated at the bottom. Some smaller particles may still remain suspended in the oil at this point, but the oil can be left to rest for two weeks or longer while this additional sediment continues to settle. When you are ready to use it, simply decant the oil, leaving this sediment behind.

At this point you could run your oil through a filter to further purify it, removing even the slightest remains from the pressing process and thus increasing the smoke point as well as the shelf life of your product. This can be as simple as pouring the oil through a fine mesh screen, or even a coffee filter placed inside a funnel. This can be a slow process but will ensure the highest quality of oil. Larger-scale production facilities

A tabletop model of a motorized oil press being used to press hempseeds.

Turbid hempseed oil (*left*) compared to settled hempseed oil (*right*).

make use of mechanized filter machines that pump the oil through a series of screens or cloths to remove any solids from the oil. Oftentimes, diatomaceous earth, made from the fossilized remains of tiny, aquatic organisms called diatoms, is used as a filter aid. This fine powder bonds to the smallest particles of sediment in the oil, which may otherwise pass through a cloth filter. Although the use of a filtering machine does result in the purest oil, this investment is typically far beyond the needs of a home-scale producer.

Once your oils are filtered, it's time for bottling and storage. How you choose to package your oils will depend on a handful of factors, including the intended uses for your oils as well as the resources available to you. The best option for storage is glass containers simply because they are easily accessible, but porcelain, ceramic and stainless steel all work well too. Plastic containers will suffice, but I shy away from recommending the use of plastic whenever possible. Opaque packaging is ideal if available and the oils should be stored out of direct light and away from heat. Remember, oils will oxidize, some quite quickly, so always keep your packaging sealed when not in use. The optimum storage requirements to ensure the longest shelf life and highest quality oil will vary depending on species. The specifics for each oilseed will be covered in Section Three. However, storing your oils in a cool, dark location is always recommended.

While consuming rancid oil may not make you sick in the short term, it is commonly thought that oxidized oils can negatively impact long-term health and therefore it is recommended that any rancid oils be disposed of. Rancid oil can be bottled and

Filtering freshly pressed hempseed oil using a cheesecloth bag.

put in the trash, or small amounts of seed or nut oils can simply be composted, although adding too much oil to the compost will slow down the decomposition process. Some recycling centers also accept discarded oils, which are then sent off to be transformed into biodiesel fuels. For the industrious homesteader, this is a project that can be done right at home!

Bottling can be as easy as decanting the oil into a new container, although if you're working with large quantities of oil and plan to fill a number of bottles at one time the use of a bottling wand will simplify the task. These can be purchased through any brew supply company, and there are a couple of different styles to choose from. Both rely on gravity, where the oil to be bottled is elevated above the bottles, typically in a customized bucket with a spout located near the bottom. The bottling wand is then attached to this spout via vinyl tubing. When the wand is pressed to the bottom of the bottle, the valve is opened and the oil will flow from the bucket, filling the bottle. The other style of wand utilizes a small spring to activate the valve, and while the choice of wand is purely a matter of preference, the spring-operated model does prove to be more difficult to keep clean. This specialized bottling equipment will save time but obviously involves a larger upfront investment and may not be necessary, especially for a start-up operation.

Through every step of the process keep your equipment clean and sanitized. This is particularly true for your oil press. After pressing is completed, disassemble and clean the parts thoroughly. A well-maintained machine will last longer, perform better and will help to avoid any potential cross-contamination. This is especially important if you are pressing seeds and nuts that are known allergens. Proper press maintenance and upkeep will lead to many enjoyable years of successful, do-it-yourself seed and nut oil extraction.

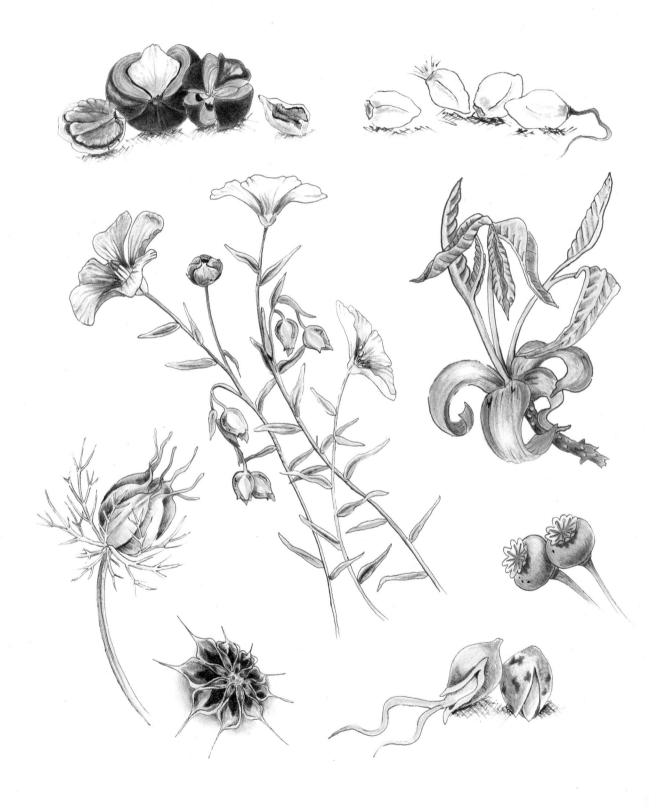

Seeds and Nuts:
The Oilseed Crops

Almond

(*Prunus dulcis*)

Almonds are an ancient crop, thought to be one of the first domesticated fruit trees, along with figs, apricots and sour cherries. In fact, almonds are closely related to a number of fruit-bearing trees, including plums, apricots, cherries and peaches, the latter of which is the almond's closest domesticated relative. Although commonly referred to as a nut, the almond we consume is actually the seed of a type of fruit known as a drupe, or stone fruit, in which the flesh of the fruit surrounds a pit that holds a single seed inside.

Wild almonds, also known as bitter almonds, are native to western Asia. These seeds contain a toxic chemical, hydrocyanic acid, and require leaching or roasting to be suitable for human consumption. At some point in history, a random genetic mutation occurred, resulting in an almond seed free of this chemical, which has since come to be known as the sweet almond. Although it is uncertain exactly when this evolution occurred, archeological evidence of sweet almonds points to somewhere between 3000–2000 BCE, and wild forms of these trees still grow along the eastern coast of the Mediterranean.

Sweet almonds were first brought to North America by the Spanish in the 16th century, but due to the cooler, wetter climate, it would be another hundred years before they could successfully be cultivated on a large enough scale to become a viable commercial crop. Although they are grown in many places in the world today, the state of California is the largest supplier of commercial almonds. While a majority of these almonds are grown for culinary use, the market for sweet almond oil has seen a dramatic increase in recent years, thanks in large part to its popular use in cosmetic products.

While bitter almonds are utilized for the extraction of essential oils, as well as to flavor liqueurs, it's the oil pressed from sweet almonds, enjoyed for its various culinary and cosmetic applications, that is of value to the homesteader and small-scale producer. Almonds yield an average of 53% oil, but depending on the cultivar this can vary anywhere from 45%–60%. The oil is a light yellow color, with a mild aroma and light, somewhat sweet flavor.

Planting and Growing

Growing almonds requires space as well as time. Almond trees may not yield fruit for five to twelve years after planting but, once mature, will likely continue to offer annual harvests for 25 years or more. Depending on the grower's available time and skills, an almond orchard can be started from seed, or young trees can be purchased from a local nursery and transplanted into the chosen area. Growing almond trees from seed is certainly the slowest method, but with some patience and attention to detail, this is still a relatively easy way to propagate these fruit trees.

Almonds must be in their whole, raw state to be as used as seed; most almonds sold in the United States and other places are pasteurized, and therefore will not grow. Soak the seeds in their shells overnight to speed up germination. Before planting, gently crack the shell open on one side, making it easier for the young sprout to escape, but do not remove the shell. Place the seeds into a small container, covering with moist peat moss or other light medium; cover with plastic wrap to help retain moisture, and place the container into a refrigerator for one to two months. Some growers suggest gently breaking the pointy tips off each almond before placing them in refrigeration. Alternatively, the soaked seeds can be added to the growing medium and placed inside a resealable sandwich bag. This process, known as stratification, will help break the seeds' dormancy and dramatically increase germination rates. After stratification, the almond seeds can be planted into potting soil, in containers, at a depth of about one inch (2.54 cm). Keep the soil evenly moist until young seedlings have emerged. Once the young trees have reached a height of 18–24 inches (46–61 cm) they can be transplanted out into the orchard. Whether transplanting saplings grown from seeds or purchased from a local nursery or other source, be sure that the soil is rich with organic material, and keep the young trees well watered until they are established.

Almond flowers are self-incompatible, meaning that more than one tree, of two different cultivars, is needed to ensure pollination and achieve fruit set. Pollen is moved from one flower to the next by bees, and maintaining a small apiary is a worthwhile consideration for the homesteader looking to grow almonds with great success. In California, the largest almond-growing region in North America, with over a million acres (404,685 hectares) of orchards, nearly two and a half million bee colonies are trucked in annually to pollinate the flowers and guarantee the harvest. Breeders have been working to develop self-pollinating almond cultivars, which wouldn't require a second tree to be able to produce fruit. A geneticist in California developed the first tree of this type in 2010, and six years later a self-fertile variety known as Independence was introduced on the market. Since then, other self-pollinating almonds have been developed and have gained popularity, especially among small-scale growers with limited space.

Mature trees will begin to set fruit as early as five years after planting, although some trees could take twice as long. The fruits resemble oblong, green-colored peaches. As they ripen, the outer hull will split, exposing the almond seed within its shell inside the hull. At this stage, the almonds are ready to be harvested. Commercial growers often use machines to shake the trees, knocking all of the mature almonds to the ground, but the small-scale grower can simply use a long stick or plastic pole to help the tree release its seeds. Larger, thicker branches can be struck with a rubber mallet. This chore can be somewhat hazardous, so consider wearing a hardhat, safety glasses or other protective gear.

Once down from the tree, the hulls must immediately be removed from the almonds in order to avoid rotting and mold growth. While commercial growers use specialized machinery for this task, small-scale almond harvesters will likely need to do this by hand. Once hulled, the almonds will require further drying and should be spread out onto screens, perhaps in front of fans, in a sheltered area. If drying outdoors, consider covering the almonds with a screen or plastic netting to protect them from hungry birds. Check the almonds occasionally for dryness by removing the shell and breaking the kernels. If the almond bends, it is still too moist, but if it snaps, then the seeds are ready to be put into storage or pressed for oil.

Post-Harvest Processing

Almonds can be stored, in shell, at room temperature for up to eight months or for over a year if refrigerated. Freezing the seeds for 48 hours before storing them will kill any potential pests or insect eggs. Since almonds easily absorb outside odors, they should be stored in airtight containers.

Almonds need to be shelled before being run through the oil press as the shells are quite hard and likely to jam up the machine. Large-scale producers use specialized shelling machinery that feeds the almonds through a series of rollers to crack and remove the shells. This can be replicated at home using a rolling pin or hammer to break the shells, or they can simply be cracked and removed using a nutcracker. Various hand-cranked nutcrackers and other hand-operated tools are available that can make this tedious chore less time consuming. Once shelled, the seeds should be separated from the debris before being run through the press.

Pressing

Shelled almonds can be pressed raw but yields will be higher with seeds that are heated before extraction. For maximum yields, crumble the almonds, either by hand or with a food processor, and lightly roast the seeds in a 375°F (190°C) oven for approximately five minutes. If purchasing almonds for oil extraction, consider the slivered or chopped options since they are typically less expensive than whole almonds and will save you from having to break the seeds yourself.

The almond seedcake is low in fiber and high in protein, and makes an excellent gluten-free baking flour when milled into a fine powder. The remnants of almond seed oil extraction could also be used as

Whole, slivered and in-shell almonds on display in a wooden bowl.

a supplemental feed for livestock, but considering the labor involved in harvesting and shelling, and the high price of purchasing, saving this sweet and flavorful presscake for personal use is the most sensible option.

Storage

Since its high vitamin E content acts as a preservative, almond oil can be stored at room temperature for up to six months in an airtight container before rancidity becomes a concern. Like all unrefined seed and nut oils, cool and dark storage is the key to longevity of the product. While any temperature below 50°F (10°C) is recommended, storing almond oil in the refrigerator is considered ideal as oil stored in this manner will remain shelf stable for over a year, even after opening.

Uses

Almond oil has a relatively high smoke point of around 428°F (220°C) and can be enjoyed in a number of culinary applications, including sautéing and stir frying, as well as grilling or baking. Although the oil can certainly handle these high temperatures, it may not always be a suitable option for your recipe since the oil will impart its sweet flavor into the dish. In very high-heat applications, the delicate flavor can be completely lost, meaning that this labor-intensive oil may not be the right choice. Almond oil is useful as a salad dressing and marinade and also makes a wonderful finishing oil, adding a sweet and nutty flavor to the dish.

A majority of the almond oil produced commercially is destined for use in cosmetics. A myriad of skin and hair care products are available utilizing the emollient and nourishing properties of this oil. High in vitamins A and E, almond oil is thought to improve the complexion, protect the skin from sun damage and prevent stretch marks, among many other benefits. In addition to this, sweet almond oil is used as a carrier oil for essential oil blends thanks in part to its nourishing effects but also for its light body and quick absorption into the skin. Oftentimes, highly refined almond oil is used in commercial products due to its lack of aroma, but cold-pressed oils can be used for any of these applications.

Brassica

(*Brassica* spp.)

Numerous brassica species have been utilized as oilseed crops, with the earliest recorded likely being *Brassica nigra*, black mustard seed, having been pressed for its oil in India as early as 2000 BCE. In China, and further north into Russia, it was the brown-seeded mustard, *B. juncea*, that was commonly grown for culinary applications as well as for its oil. It is also quite possible that wild field mustard, *Brassica arvensis* or *Sinapis arvensis*, was an early oilseed crop in its native regions of northern Africa and Southeast Asia.

Today, the most common brassica seed oil is rapeseed, *B. napus*, which is sold commercially under the name canola oil. This particular plant is the third largest source of vegetable oil in the world. Closely related to the rutabaga, rapeseed didn't see widespread commercial success until the 1970s, after Canadian agricultural scientists and the corporations they worked for launched a massive marketing campaign to promote their newly developed cultivars: rapeseed varieties that yielded oils low in erucic acid. Until these new strains were developed, traditional rapeseed oil was considered unsafe for human or animal consumption due to its high levels of erucic acid, an organic compound believed to cause damage to the heart and cardiovascular system. Of course, savvy marketers realized that a product called rapeseed oil may not appeal to consumers, and so this new product was dubbed "canola," a contraction of the words "Canada oil, low acid." The modern cultivars are considered food safe, and stringent government oversight ensures that the erucic acid content of these varieties remains below 2% by weight. In the late 1990s a genetically modified variant of rapeseed was developed to be resistant to herbicide applications, and within two decades these new GMO varieties have dominated the commercial market. Unrefined oil extracted from food-quality rapeseed is a vibrant yellow color and exhibits a mild, nutty flavor.

In addition to mustard and rapeseed oil, other members of the Brassicaceae family also offer seeds worth consideration for oil production. Among these is radish, *Raphanus sativus*, which produces an oil renowned by the health food community for its nutritional benefits and which is often utilized in commercial skin and hair care products. While any radish can be used as an oilseed crop, particular varietals have been developed for this specific purpose. These radishes are similar in appearance to daikon radishes, with their long, tapered roots, and are often used in similar ways. Oilseed radishes are often referred to as var. *oleiferus*, and many of them double as forage or cover crops to help loosen and nourish compacted soils. Radish seed oil has very little scent and is a light yellow to golden color.

Planting and Growing

There are two distinct types of rapeseed that can be grown, annual spring-planted varietals and biennial, fall-planted types. The latter requires vernalization, or overwintering, in order to trigger flowering and is the more popular of the two rapeseeds amongst commercial growers. Not only do winter types yield 20%–30% more than spring-planted canola, but since they are typically planted in September and harvested by June, these cool-weather crops are able to avoid the hottest weather of the year.

Regardless of when the rapeseed is planted, the technique is the same: simply broadcast the small seeds onto the plot. The plants are tolerant of most soil types as long as the area is well-draining and in full sun. If you are row planting, seeds should be spaced around three inches (7.6 cm) in rows about seven inches (18 cm) apart.

After flowering, the seed pods, known as siliques, will begin to form. Canola is an indeterminate crop, meaning that the seeds will mature at varying rates. Often, commercial farmers will swath their canola fields when the bottom third of the seed pods on the plant stems have reached maturity. Swathing is a common technique in cereal grain farming, in which the plants are cut just below the pods and then stacked upon the field stubble to finish drying down. This technique is used to avoid seed loss from shattering of the earliest maturing pods while also giving the youngest seed pods more time to ripen, thus increasing total yields. Swathed fields are typically left to rest for around ten days before being combined.

The small-scale grower can consider similar methods for harvesting their rapeseed crop. Canola stems can be cut at the same stage of maturity, when the bottom third of the pods have dried and the seeds inside are a dark brown or black color, and laid out to finish drying either in the field or on screens or tarps in a separate location. Leaving brassica pods to finish maturing in a small hoop house or polytunnel is a common practice. Once dried, threshing the seeds is a simple task as the pods easily shatter under the slightest impact. Rapeseed can be winnowed to remove any debris by use of a fan, but running the seed through a series of screens may be the easiest and most efficient method. Often the small seeds of brassicas can be blown away along with the seed pod debris during winnowing, so care must be taken to avoid this.

Radishes grown as oilseed crops are often fall-planted, over-wintering daikon types, although spring radishes can also be cultivated for this purpose. As members of the Brassicaceae family, radish seeds are also borne in siliques along sturdy stems, yet these pods are shatter resistant and require manual threshing. Once collected, the seeds can be processed and cleaned in a fashion similar to canola and other brassicas.

Post-Harvest Processing

Properly cleaned and well-dried brassica seed can be stored in airtight, food-grade containers for months before pressing. While a majority of commercial canola oil is extracted under high heat and by the use of hexane as a chemical solvent, there has been increased interest among the health food community in expeller-pressed canola oils. Although this method of production offers smaller yields, and therefore higher prices, it is widely thought that these chemical-free alternatives are superior in flavor and nutritional value when compared to the mass-produced, ultra-refined products.

Canola seeds can certainly be heated before being run through the oil press, which will typically increase yields by up to ten percent. A temperature of no more than 120°F (49°C) is ideal for preparing brassica seed for oil extraction.

Pressing

The average oil content of canola seed is around 40%. Expeller pressing preheated seeds will pull about 85% of the available oil, while cold pressing will yield notably less, sometimes as little as 65%.

Seeds can be loaded into the hopper and pressed much like any other seed. The resulting seedcake is ideal for livestock feed supplement due to its high protein content and is one of the most widely used feed supplements for commercial cattle. If you are cold pressing canola seeds, consider running the seedcake through the press a second time to increase production.

Storage

Brassica seed oil is quite shelf stable, and an unopened container can be stored at room temperature, away from heat and light, for at least six months. The optimum storage temperature for cold-pressed canola oil is around 39°F (4°C), and refrigerated oil will keep for up to a year even after opening. Cold-pressed oils are more susceptible to early rancidity than oils extracted under high temperatures, so care should be taken to keep the oil fresh.

Uses

Unrefined, cold-pressed canola oil has a smoke point of around 374°F (190°C), and after thorough filtering, the oil can be heated to 399°F (204°C) before smoking occurs. Expeller-pressed canola oil can handle temperatures up to 450°F (232°C). This means that canola oil is ideal for high-temperature cooking, including frying and sautéing. Its light flavor also makes the oil ideal for baking or even as a marinade or salad dressing oil.

Due to its high content of essential fatty acids and vitamin K, canola oil is also an ideal choice for use in skin care products. The oil is thought to be anti-inflammatory, and commercial products for the treatment of eczema often include canola oil among their ingredients. Numerous moisturizing lotions containing rapeseed oil are available on the market as well.

Radish seed oil has a long history of use in skin care products, dating back to ancient Rome. Useful as a skin moisturizer and believed to neutralize free radical damage, radish seed oil can be found in a myriad of products, yet the most common cosmetic use for this oil is to nourish and maintain healthy hair, scalp and nails.

In addition to these cosmetic and culinary uses, brassica seed oils are utilized around the world in the production of biodiesel. This is due to the seed's high oil yields as well as its low saturated fat content which produces a more efficient fuel product. The industrious homesteader could certainly consider brassica oils as a component in small-scale biofuel production.

Flax   (*Linum usitatissimum*)

Flaxseed is one of the oldest cultivated crops, and its use in the production of linen dates back to ancient Egypt, if not earlier. While there are no known wild populations of *L. usitatissimum,* the closest wild relative, *Linum bienne,* is native to the Mediterranean region and western Europe. Archeological evidence of domesticated oilseed flax cultivation dating back nearly 5,000 years has been found in the area now known as Syria.

Very early in the domestication process, two distinct variations of flax were developed, a larger-seeded, oilseed type and a fiber crop variety used in the production of linen. The common culinary flax, with either brown- or blond-colored seeds is preferred for oil pressing. While chefs may have a preference for which color of seed they choose to work with, nutritionally the seeds are similar and will produce oils of equal value.

In the early 1990s, a new cultivar of flax was developed under the trademark name Linola. This new varietal was bred to be low in omega-3 fatty acids and marketed as a livestock feed. Although the high omega-3 content of flaxseed is an attribute celebrated by most for its believed health benefits, this high alpha-linolenic acid content also dramatically shortens the shelf life of the crushed seeds and oil. This new cultivar, Linola, had improved storage qualities due to the lower ALA and is considered superior as a bulk livestock feed.

Outside of the health food store, the oil of flaxseed is often sold under the name linseed oil. This oil is used as a varnish on wood, as an ingredient in paint and also in some soaps, inks and even in the production of linoleum. Flaxseed oil pressed at home can be used for any of these applications, and caution should be taken if purchasing commercially distributed linseed oil. Oftentimes, these products are solvent-extracted oils and in some instances are blended with polymerized oils, industrial solvents and/or metallic dryers such as cobalt and manganese.

Pure flaxseed oil is considered one of the healthiest oils that one can consume. In addition to the omega-3 fatty acids previously mentioned, flaxseed oil is high in B vitamins and minerals such as magnesium and iron. It's often enjoyed as a nutritional supplement, either in capsules or added to smoothies but can also be used as a substitute for other oils in salad dressings, sauces or other recipes.

Flaxseed oil is a golden-yellow color regardless of which color of seed is pressed, although brown seeds may produce a slightly darker oil. Regardless, these oils are nutritionally equal and can be used interchangeably. The oil should not be used for cooking, as the heat will degrade its quality, but can be utilized in unheated culinary applications as well as topically.

Commercial skin creams make use of flaxseed's many health benefits, and producers promote their products as beneficial in the treatment of psoriasis and eczema as well as to smooth wrinkles and soften aged skin.

Planting and Growing

Flax takes around 110 days to produce a mature seed crop and should be planted in the early spring, although some varietals will mature in as few as 90 days. Flax thrives in cool weather, so planting early is important. Seeds should be sown around the same time one would typically plant peas, lettuce and other cool-weather crops. Direct sowing of seeds is recommended, and the seed bed should be kept moist until germination, about ten days. Flax will perform best in rich, well-drained soils, and it's important to keep weeds under control until the plants are well established. In areas with a very short growing season, flax can be started indoors and transplanted out into the field as soon as the soil can be worked.

Harvesting

Plants are ready to be harvested when a majority of the seed pods have turned from green to golden. The seeds will rattle inside the pod, signaling their maturity. On a large scale, flaxseed can then be harvested using a combine, but in a smaller plot they can easily be harvested by hand. The quickest way to gather flax by hand is using a scythe, but seed heads can be removed with scissors, or entire plants can be pulled from the earth and bundled.

If seed heads are harvested by hand, they will then need to be threshed and winnowed. The size of the harvest will determine the threshing technique—whether the seed heads are crushed by hand using a flail or other tool, or even stomped underfoot, thus breaking open the pods and releasing the seeds. Cleaning the chaff from the seeds can be accomplished using screens to sift out the debris or by gently pouring the seeds and chaff in front of a small fan to blow the plant material clear while the seeds themselves drop into a bucket or onto a tarp below.

The ideal moisture content for oil extraction is 7%–11%, and at this point, the freshly winnowed seeds are ready for pressing.

Post-Harvest Processing

Flaxseeds can be pressed whole and will not need to be shelled. There is also no need to heat the seeds before running them through the press. In fact, exposing flaxseeds to high temperatures, over 120°F (48°C), will negatively affect the quality of the final product.

It's important to note that due to the high omega-3 content of flax-seeds, the cracked seed as well as the oil will oxidize, quickly turning the product rancid. If you choose to crush your seeds before running them through your press, only crush the quantity that is needed for that session in order to maintain the highest quality. All crushed flax-seeds, oils, and any seedcake that will be retained and used should be refrigerated immediately.

Pressing

Both brown and blonde flaxseeds can be pressed for their oil.

Whole flaxseeds are very hard and can be challenging to press when using a manual, turn-screw-style expeller press, especially the initial cranks to get started. A modified, bicycle-powered unit, as described in

an earlier chapter, will certainly make the job easier, as would a motorized machine.

Once the oil is extracted, it will need to settle before decanting. Be sure to refrigerate the oil to maintain its quality. Let the oil sit, undisturbed, for up to a week to improve its clarity. If desired, the decanted oil can then be filtered through a cheesecloth or mesh screen to remove even the most minute debris.

Storage

A mentioned above, flax oil should be stored in the refrigerator for maximum longevity. The optimum temperature range for storage is 37°–45°F (3°–7°C). An unopened, opaque bottle of flaxseed oil will keep its quality for up to a year when stored this way. Once the bottle has been opened for use, the shelf life drops to a short six to eight weeks. One of the many benefits of pressing oil at home is always knowing the expiration date of the product and being able to produce small batches of fresh oil whenever you need it.

The seedcake resulting from flaxseed oil production can be milled for use in the home kitchen but should also be kept cool in storage. A package of crushed flaxseed can be stored for up to three months in the refrigerator and six to eight months when kept in the freezer.

Uses

Flax oil is one of the most delicate culinary oils, with a limited shelf life and a very low smoke point of 225°F (107°C). In addition to the many applications noted earlier, the oil is highly recommended for use in seasoning cast iron cookware, since this low smoke point means that the oil is quick to polymerize, creating a hard, stick-resistant surface that protects the metal from both air and food.

The flax seedcake can be easily milled into a fine powder using a household food processor. This powder can be incorporated into baked goods, added to smoothies or even used to top cold or hot breakfast cereals. A tablespoon of ground flaxseed mixed with three tablespoons of cold water can be used to replace an egg in any recipe. Just thoroughly mix the finely ground seed and water and let the mixture rest in the refrigerator for 10–15 minutes to thicken before using.

Grapeseed

(*Vitis vinifera*)

Clusters of grapes grow on what are known as lianas: long-stemmed, woody vines that are rooted in the soil at ground level while utilizing trees or other means for vertical support. There are more than 45 described species of *Vitis*, most of which are isolated to the northern hemisphere, and the vine most well known for wine production is native to central Europe, the Mediterranean and southwest Asia. Domesticated grape vines have hermaphroditic, perfect flowers, while wild types are mostly dioecious, meaning that male and female flowers appear on separate plants. Wild grapes have long been harvested as a source of wild food and medicine, and the first domestication of these flavorful berries is believed to have taken place around 3500 BCE, near the Black Sea in the area now known as the country of Georgia.

Winemaking accounts for approximately 90% of all modern grape cultivation worldwide, and even in ancient times the production of wine was the most common use for these fruits. In ancient Greece it was Dionysus, the son of Zeus, who was credited with introducing this intoxicating beverage to the world. Great celebrations, which involved plenty of imbibing, were thrown in his honor. The Roman adaptation of this god was known as Bacchus, and his festivals were called Bacchanalia, a term still used to this day to describe wild celebrations of overabundance.

Although grapes have played a significant role in society for thousands of years, the oil pressed from grapeseed is a relatively new product. Grapeseed oil was first pressed in France in the early 1930s and didn't see worldwide popularity until the mid-90s. Grapeseed oil is prized for its neutral, clean taste and high smoke point, making it a welcome culinary oil with a wide range of applications.

Most, if not all, commercially available grapeseed oil is produced from seeds collected from the byproduct of the wine industry. After the grapes are crushed and fermented, the remaining pomace, also known as marc, which contains any solids remaining after pressing, including the seeds, skins and pulp, is then washed and cleaned through screens. This separates the seeds from the rest of the product. The seeds are then dried to a moisture level of around 8%, at which point they are ready to be pressed for oil. Some modern wineries are experimenting with a process called *delestage*, also known as rack and return, which removes the seeds from the must during the fermentation process. It's thought that this technique may have a favorable effect on the flavor of the wine, while not seeming to have any impact on the quality of the oil extracted from these particular seeds.

In North America, wine has also been made from the grapes of other native species including *Vitis labrusca*, *V. rotundifolia* and *V. riparia*. It's certainly possible for the industrious oil presser to work with these species as well, although finding a source of bulk seed may prove difficult as these wine products are far less widely manufactured. It's important to note that extracting grapeseed can prove to be labor intensive and difficult, especially when working on a small-scale operation. The typical oil content of the grapeseed is usually anywhere from 14%–20%, although black-seeded types are notably lower, sometimes as low as 6% oil.

Planting and Growing

Growing one's own grapes is a multi-year commitment of time and space. Well-cared-for vines will typically begin to produce within the first few years and will continue to bear fruit for 30 years or more. The key to successful production is proper pruning, as the fruits produce only on new growth. Maintaining a healthy and prolific vineyard is an art, but it can be accomplished by anyone willing to take the time to learn the growth habit of the plant and the techniques required to keep the vines organized, pruned and trellised.

There are three main types of grapes to select from for home production: *V. labrusca*, which tends to be a more cold-hardy species, *V. vinifera*, which is best for wine production and thrives in warmer, dryer climates and hybrids of these two species. Hybrids are certainly worth consideration as they tend to be cold tolerant and disease resistant, although it is commonly thought that the fruit of hybrid grapes is not as flavorful as their European relatives.

One-year-old vines can be purchased from a reputable nursery and should be planted in the early spring in a full sun area. Space plants at least six feet (1.8 m) apart and plan to construct a sturdy trellis or arbor to support the growing vines. Many growers suggest soaking the roots in water for up to three hours before planting and then watering the plants deeply once transplanted. Begin a proper pruning regimen beginning the second year and you should be rewarded with your first grape harvest by the third growing season.

Post-Harvest Processing

Extracting seeds from vine-ripened grapes can be a slow and laborious process. These seeds are typically collected from the byproduct of winemaking and for the home producer this is a logical path to follow.

Once the grapes have been crushed and put through the fermentation process the resulting pomace can be spread onto screens and washed with a high-pressure water application to further separate the seeds from the skin and pulp of the fruits. These seeds can then be spread onto drying racks, preferably in front of fans, and left to dry.

Due to the amount of effort required to collect grapeseeds, and their relatively low oil content when compared to other oilseed crops, you might want to consider simply sourcing grapeseeds from a local winemaker or through an online supplier, especially when first starting out as a small-scale oil producer. Connecting with a local winemaker is a great option if you are lucky enough to have one in your area; this would be a mutually beneficial relationship. You can keep your supply chain local while your winemaker finds a profitable use for their byproduct.

Pressing

A great majority of the commercial grapeseed oil available on the market is produced by chemical extraction through the use of hexane. This is due to the low oil content of the grapeseed and the higher productivity, and therefore profitability, of extracting the oil via these modern methods. Cold pressing, and even high-heat mechanical pressing, offer lower yields, and this is reflected in the notably higher prices for these particular products. Cold pressing grapeseed produces a deep golden oil but will require large quantities of seeds for a modest return and may not be feasible for your operation. Heating the seeds before pressing will notably increase oil yields. The recommended temperature to heat seeds for maximum oil production is 200°F (93°C). This high-heat extraction will produce a dark green, almost brown-colored oil that will maintain a slighter higher smoke point than the cold-pressed oil, even when the latter is well-filtered. Grinding the grapeseeds before running them through the press may also help increase their productivity and usefulness as an oilseed crop.

The resulting seedcake can be used in the same ways one might use the ground grapeseed products available through numerous health food outlets. These products are added to protein powder mixes, foods and drinks as a nutritional supplement and for their high fiber content.

Beautiful Marquette grapes ripening on the vine at a winery in Michigan.

Storage

Like other oils, grapeseed oil should be stored in airtight containers away from heat and light. The use of opaque packaging is ideal for extending the shelf life of the oil. Due to its naturally high levels of vitamin E, grapeseed oil can be stored at room temperatures of 68°–77°F (20°–25°C) for six to ten months before rancidity becomes a concern. An unopened bottle will remain shelf stable at these temperatures for twice as long. Oil produced through high-heat extraction will remain shelf stable for longer than cold-pressed oils. Once opened, it's recommended that cold-pressed grapeseed oil be stored in the refrigerator where it will keep for up to six months.

Uses

Grapeseed is a versatile oil than can be used for a wide range of culinary applications. Thanks to its neutral flavor and high smoke point, this oil is an excellent choice for sautéing, stir frying and other high-heat cooking. The unrefined, cold-pressed oil has a smoke point of 380°F (193°C), while oil pressed from seeds that were heated before extraction can be used at temperatures up to 400°F (204°C) before smoking becomes an issue. Although the cold-pressed oil is able to handle such temperatures, this heat will still break down much of the nutritional value of this product, and therefore it's recommended to use this particular oil for cold culinary purposes such as salad dressings, marinades and as a finishing oil, drizzled over dishes immediately before serving.

The oil's light scent and high vitamin E content also make it an ideal selection for use in skin care products and as a carrier oil for use with essential oils, which are typically steam-distilled from select botanicals. Since grapeseed oil is considered antimicrobial, it is thought to be useful in treating acne and is often used by herbalists in formulations to moisturize and soften aged skin, as well as for its anti-inflammatory effect.

Hazelnut

(Corylus spp.)

A member of the birch family, the hazel genus *Corylus* is composed of around fourteen to eighteen distinct species, all of which produce edible seeds commonly referred to as hazelnuts. Only two of these species are cultivated commercially, *C. avella*, the common hazelnut and *C. maxima*, also known as the filbert. Both of these trees are native to Europe and western Asia but today are grown around the world for their edible seeds, with Turkish hazelnut production accounting for a majority of the world supply. These particular species are considered superior for nut production on any large scale for a number of reasons including the larger size of the seeds as well as the thinner shells, which makes cracking the nuts less labor intensive. In addition to this, some species such as beaked hazelnuts, *C. cornuta*, have spiny involucres, or leaf bract, which surround the nut, making the chore of shelling more difficult. American hazelnuts do not easily drop from the plant when mature, unlike their cultivated cousins, and therefore need to be picked by hand. They will often also need to have the involucres removed before curing, adding yet another step to the already laborious process. For small-scale hazelnut oil production, wild types are certainly valuable, but the advantages of working with domesticated species are worth consideration.

People have no doubt utilized the nuts of hazel trees as a food source for thousands of years and the earliest archeological evidence of hazelnut oil production can be traced back to northern Italy, sometime during the Bronze Age. At the time, the oil was extracted simply by crushing the nuts between two millstones, and this process is somewhat replicated through the use of an expeller press today. Hazelnut oil saw a cultural resurgence in Italy during the Second World War. With rations making many food items scarce, including cooking oils, ingenious local farmers began producing hazelnut oil from trees on their homesteads to supplement the supply. The flavorful oil added an element of luxury to an otherwise simple meal and hazelnut oil is now considered an essential ingredient in many traditional Italian dishes.

Planting, Growing and Foraging

The work of foraging hazelnuts begins anywhere from late summer to early fall depending on your location. Locating and identifying the trees in advance of harvest season would be practical. Nuts can be harvested from the trees, or shrubs in many cases, when the shells have begun to brown yet the involucres are still green. For European hazelnuts, the mature nuts will often drop to the ground and should be collected as quickly as possible to prevent mold, worms or other critters getting to them first. Hazelnuts still encased in their leaf bracts should be removed. If removal is difficult, spread the nuts out to dry for a few days, or until the involucres turn brown and then remove them at this time. When working with species with spiny husks, wear gloves to protect your hands. Spread the nut, in shell, out to cure for one to two weeks in a dry area out of direct sunlight. Properly cured hazelnuts will keep in their shells for up to a year if stored in a cool dark location. Airtight, food-grade buckets will protect the nuts from rodents and refrigeration will double their shelf life.

Hazelnuts can be grown from seed, but purchasing established trees from a local nursery is recommended. Since the seeds are produced through cross-pollination of two cultivars, it is difficult to predict the quality of nuts grown from seed. This cross-pollination requirement also means that you will need at least two hazel trees of different cultivars to ensure nut formation. Hazel trees are monoecious, meaning they have separate male and female flowers on the same tree, but these flowers rarely bloom at the same time, making the second tree a necessity. Hazel trees prefer full sun and will typically begin producing nuts anywhere from two to five years after planting, although the initial harvests may be small.

Post-Harvest Processing

Once dried, the hazelnuts will need to be shelled before they can be run through the oil press. For very small batches, this can be accomplished with a hand-operated nutcracker but for larger-scale production a tabletop, lever-action cracker will work well. For a small investment, consider purchasing a hand-cranked nutcracker tool. These machines can be adjusted to accommodate nuts of various sizes that are loaded into a hopper and moved through the cracker by a hand-turned crank. Electric models are also available but with some basic modifications, the hand-operated machine could be upgraded to include a motor or even rigged to a bicycle to utilize pedal power. The nuts will run through the press more smoothly if broken or crushed, so any whole nuts should be broken down after cracking, either through a food processor or crushed with a tool at the bottom of a bucket or other container.

Pressing

After being shelled and sorted, the nuts can be run through the oil press raw or, to produce an oil with a deep nutty flavor, they can be roasted before the extraction process. Simply bake the nuts in a 350°F (176°C) oven for 10–15 minutes, while stirring occasionally to keep them from burning. Let the nuts cool for three to five minutes after removing them from the oven before loading them into the oil press. Hazelnuts range from 57%–65% oil depending on cultivar. Roasting the nuts before extraction will maximize the potential yields of the variety pressed while adding an additional layer of flavor to the final product.

The seedcake created by the hazelnut oil extraction process is high in both protein and fiber. It can be milled into a powder to be used in smoothies, shakes and cereals or in gluten-free breads and pastries.

Storage

An unopened bottle of hazelnut oil is considered shelf stable at room temperatures for up to six months. Once opened, the product should be refrigerated to avoid rancidity. Stored at temperatures below 45°F (7°C), hazelnut oil will easily keep for a year and once opened, will remain fresh for up to eight months. Like all other oils, hazelnut oil should be stored in an airtight, opaque container away from direct heat and light.

Uses

Hazelnut oil is a versatile culinary ingredient that can be enjoyed in a wide range of applications. With a smoke point of 430°F (220°C) the oil can certainly be used for high-heat applications such as sauté, baking, roasting and grilling. It's important to consider that the oil's unique nutty flavor will come through in whatever is being cooked, which is why many chefs choose a more neutral oil for many of these techniques. Hazelnut oil is also an excellent choice for cold applications such as salad dressing, marinades and bread dips. Many recipes call for the use of hazelnut oil as a finishing oil, or as a substitute for butter, drizzled over pastas, grilled or roasted vegetables and grain dishes.

Herbalists often use hazelnut oil as a carrier oil for massage blends or to dilute essential oils, although the rich nutty aroma of hazelnut is sure to carry over into the products. Hazelnuts are high in tannins which gives the oil an astringent quality making it useful for treating oily skin, cleansing pores and clearing up acne. Its high vitamin E content also makes the oil useful as a moisturizer, to reduce the appearance of scars and to protect the skin from sun damage.

A hopper full of hazelnuts waiting to be shelled in a crank-operated nutcracker.

Hempseed

(*Cannabis sativa*)

Hemp is one of the oldest cultivated plants used for textiles and cordage, with evidence pointing to central and western Asia as the center of domestication. Archeologists have uncovered remnants of hemp cloth in ancient Mesopotamia dating back nearly 9,000 years, and written records mention the cultivation of hemp in China as early as 2800 BCE. An ancient Chinese herbal text known as "Pen Ts'ao Ching," believed to have been written by the legendary Emperor Shennong in 2737 BCE, is the first recorded evidence of the production and use of hempseed oil.

How hemp eventually reached North America is a matter that is still up for debate. The most widely accepted belief is that hempseed was first brought overseas by the English to what became the colony of Jamestown. Ship records indicate that hemp was included in supplies brought to the colony in 1611, and eight years later the colony established a law requiring all residents to grow the plant for the fabrication of rope and fabric. Others believe that hemp may have already grown on the continent before the colonists' arrival. In the 16th century, French explorer Jacques Cartier wrote that the land was "frill of hempe which groweth of itselfe, which is as good as possibly may be scene, and as strong," but it is quite possible that Cartier was referring to any number of native plants being used for cordage, such as stinging nettles, *Urtica dioica*.

Regardless of how hemp made its way around the world, the oil pressed from its seeds has been valued by countless cultures for centuries for its many culinary, medicinal and even industrial applications. Hempseed oil is not to be confused with hash oil, a psychoactive product derived from the cannabis flower. The principle psychoactive component is tetrahydrocannabinol, or THC, but hempseed is pressed from plants that have been bred to contain little to none of this compound. There is another product on the market made from hemp plants that has gained tremendous popularity in recent years, commonly referred to as CBD oil. This product is made by extracting cannabidiol from the stems, stalks, leaves and flowers of the plant and is quite different from hempseed oil, although pure hempseed oil is often used as a carrier oil for the CBD products.

Cold-pressed hempseed oil is a deep green color, due in part to the high level of chlorophylls found in the seeds, and has a light, nutty flavor. The oil is high in unsaturated fats and has a limited shelf life, so pressing small batches as needed is ideal. The 3:1 ratio of omega-6 to omega-3 fatty acids found in hempseed oil has made it one of the most popular health food products on the market. It is widely utilized as a nutritional supplement as well as in numerous health and beauty products. While hempseed oil also has industrial applications, such as paints, fuels and plastics, these typically utilize a highly refined, bleached and deodorized oil.

Planting and Growing

Hemp is a fast-growing plant that will thrive in most soil types and can produce a seed crop in 100–150, days depending upon the cultivar.

Plants require full sun and well-draining soil. Seeds should be sown half an inch to an inch (2.5 cm) deep and at least four inches (10 cm) apart, although giving plants more space will encourage branching and increase yields. Seedlings require consistent watering for the first six weeks, but once established the plants are quite drought tolerant. Hemp plants are what is known as dioecious, meaning that some plants contain exclusively male pollen-shedding flowers while others have entirely female pollen-receptive flowers. A decent population of dioecious plants is needed in order to ensure adequate pollination and, therefore, reliable seed set.

Plants will signal maturity when the leaves on the seed heads (found on the female plants) begin to yellow. It's time to harvest when the stems of the plants have shed a majority of their leaves and the seeds at the base of the leaf stem, on the bottom of the seed head, have turned grey. Larger-scale growers can run combines through their field to harvest and clean their hempseed, while smaller plots can be cut and gathered by hand. Simply cut the stems below the lowest seed head with a sickle or shears. Hempseeds can quickly be threshed by laying the seed heads out onto a tarp and using a flail to knock the seeds free from the plants or by beating the stems along the inside of a bucket, dislodging the seeds, which are then collected at the bottom of the bucket.

The threshed seeds will then need to be winnowed to remove any additional plant material and debris from the harvest. This can easily be accomplished by simply pouring the seeds from one container into another while allowing the wind, or a fan, to blow the chaff away. The seeds can be further cleaned by sifting them through a series of screens.

It's important to note that in many places of the world hemp production is restricted or regulated in some form or another. In some countries it's illegal to farm hemp, while other countries may require licensing, permits or even testing of the crop to verify its THC content. In the U.S., each state may have different requirements or laws regulating hemp production. Be sure to check with all regulatory bodies in your area before growing your hempseed. Purchasing hempseed for oil extraction from a reputable source is also a viable option, especially for those with limited space or regulatory burdens.

Post-Harvest Processing

Well-cleaned hempseeds can be stored in food-grade buckets or other sealable containers and stored for up to a year. The ideal storage temperature is between 32°–40°F (0°–4°C). The seeds require no additional processing before pressing; they can be left intact, with their shells on. The seeds should not be heated before pressing, although if kept in cold storage, they should be allowed time to warm up to room temperature before running them through the oil press.

Pressing

The ideal temperature for pressing hempseed, to avoid degrading its nutritional value, flavor or shelf life, is 104°–140°F (40°–60°C). This can easily be accomplished with an expeller press, as long as the crankshaft isn't turned at too fast a rate. A manual oil press, upgraded through attachment to a bicycle or similar equipment, must be turned at a moderate pace to avoid the high temperatures that may result from the friction generated by overly accelerated revolutions.

The seedcake produced by the extraction process can be used as a nutritious, high-protein feed supplement for livestock, but it is also ideal for human consumption. The crushed seeds can be further milled into a powder and used for baking, in smoothies or numerous other applications.

Hemp plants growing under lights at an indoor growing facility.

Storage

Like other oils, especially those prone to early rancidity, hempseed oil should be stored in airtight, opaque containers. A sealed, unopened bottle will keep for up to a year when stored in a cool dark location, such as a pantry or cupboard. Once opened, the bottle will keep for only one to three months before turning rancid. Hempseed stored in the refrigerator will last twice as long; an unopened bottle will remain useable for up to two years; once opened it can be kept for six months before rancidity becomes an issue. Some hempseed oil manufacturers recommend storing the oil in the freezer to delay rancidity and extend its shelf life even further. While hempseed oil stored at these temperatures can become cloudy, it will remain liquid and useable.

Uses

The smoke point for unrefined hempseed oil is 330°F (176°C), so it can be used for some low-temperature cooking applications, although it is widely recommended to avoid exposing the oil to much heat at all to avoid breaking down its potent nutritional value. Cold culinary applications well suited to hempseed oil include blending into a vinaigrette, homemade mayonnaise or even as an ingredient in pesto. These uses keep the oil's nutritional value intact while making the most of its light, nutty flavor.

Hempseed oil can be found in a plethora of topical health and wellness products, although many of these utilize refined hemp oils, since the bleaching and deodorizing process creates a light, colorless and odorless product preferred by cosmetic manufacturers. Cold-pressed hempseed oil can be used to craft topical products for home use and is considered ideal for use on all skin types. Highly moisturizing, the oil can be added to lotion and salve recipes. Hempseed is thought to be beneficial in the treatment of acne, eczema and psoriasis. The oil is considered to be a humectant, which mean it draws moisture to the skin, and it can be used as a scalp treatment for dandruff, dry scalp and damaged hair. The linoleic acid and oleic acids found in hempseed oil play a crucial role in skin health and are important nutrients to help reduce fine lines and wrinkles as well as prevent signs of aging.

Hickory

(*Carya* spp.)

Although various species of hickory have grown across the northern hemisphere since antiquity, wild species can now be found only in their native habitats in North America and eastern Asia. Hickory is from the same family as walnut, *Juglanaceae*, and therefore produces its seeds in similar fashion, within small fruits known as drupes. Although a number of *Carya* species are cultivated for their wood, only one species, *C. illinoinensis*, is grown commercially for its edible nut. This species is more commonly known as the pecan and is grown in orchards throughout the southern United States and Mexico. Pecans are one of the world's most recently domesticated crops, with commercial production not beginning until the late 1880s, with the development of grafting techniques allowing for precise varietal selections. The first pecan farm wasn't established in the United States until 1932. Until this time, all pecans, as well as other hickory nuts, were simply gathered from the wild. One of the reasons that pecans became a cultivated food crop, while other hickories did not, is their relatively thin shell, making these nuts far easier to crack than other *Carya* species.

While most hickory seeds are edible, some are quite bitter and many have exceedingly difficult to crack shells and very small seeds, requiring more effort to collect in the quantities needed for oil pressing than they may be worth. In North America, there are four species considered to be the best selections for oil production: *C. illinoinensis*, the pecan; *Carya ovata*, the shagbark hickory; *C. laciniosa*, also known as shellbark; and the mockernut, *Carya alba*. The seed of this last species is small and difficult to crack, but the sweet flavor of the nut provides an enjoyable oil to those willing to make the effort. Two Asian hickory species whose seeds are collected and pressed for their oil are *Carya cathayensis* from China and *C. tonkinensis*, which is commonly referred to as the Vietnam hickory. All of these various species have their pros and cons as oilseed crops, but one of the deciding factors for the prospective producer is certainly whichever hickory is local as well as the amount of time and effort that might go into foraging and processing the nuts themselves.

Planting, Growing and Foraging

The pecan is the only hickory species in widespread cultivation for its edible nut and is an excellent option for producers who reside in the warm, humid climates the trees prefer. The trees can be grown from seed, which is a slow process taking 15 years or more from planting until the first harvest, or from grafted rootstock that could begin producing as early as five years after transplant. The seeds can easily be germinated by first stratifying the pecans for two to three months in order to break their dormancy and then planting them in a deep pot. Keep the pecan seeds moist until they sprout, which will take approximately one month. After the first season of growth, the sapling can be moved into rich, well-composted soil. Healthy young trees can also be purchased from a local nursery, and this is likely the most time-effective method for establishing a productive pecan orchard. Pecan trees are monecious, meaning that they have separate male and female flowers on the same tree, but since flowers are not pollen receptive at the same time that they are pollen shedding, a second pecan tree is needed in order to achieve pollination and fruit set. This is another good reason why purchasing established trees is beneficial, as pecans grown from seed can be unpredictable in their pollination timing as well as their nut quality.

An alternative to cultivating trees is to forage for nuts in already established, wild stands of hickory. Mature, ready-to-harvest nuts will easily drop from the trees, either on their own or with a bit of coaxing from a long stick or pole. While the domesticated pecan will often fall from the tree leaving the hull behind, other *Carya* species will need to be hulled after harvesting. This chore should be done immediately to lessen chances of mold or rot, but since the hulls tend to split upon maturity, this can be a relatively simple task to accomplish by hand. If the hulls are still green and prove difficult to remove, allowing them to sit and dry for a few days will help. Once the nuts are dried the hulls should split apart and at this time can easily be removed.

Post-Harvest Processing

Once the nuts have been harvested and hulled, they should be spread out in a single layer, in an area out of direct sunlight, to properly cure. Use fans to provide circulating air around the nuts to expedite the curing process, which could take anywhere from two to ten days. Once well dried, the seed kernels will be brittle and can easily be separated from the shell exterior. Check the hickory nuts daily to gauge their readiness. Properly cured hickory nuts can be stored in-shell and will keep for a year or more when kept at temperatures from 32°–45°F (0°–7°C). Once out of their shells, the shelf life of hickory and pecan nuts drops by half unless they are stored in a freezer, in which case they will keep for up to a year before rancidity becomes a concern.

Shelling hickories in quantities large enough for oil extraction will take some time and effort, but this task can be made easier by either freezing the nuts, which makes the shells more brittle, or by boiling the hickory seeds to soften their shells. Use of a lever-action cracker will also hasten the task. This style of nutcracker is ideal for hickory, walnuts, almonds and Brazil nuts. This may be slow work, but with some practice, a useful quantity of shelled nuts can be accumulated within a reasonable amount of time.

Pressing

Once the nuts are shelled and dried, they are ready to be run through the press. Hickory and pecan can be pressed at room temperature, which will produce a light, delicately flavored oil, or can be roasted to bring out a deep nutty flavor prized by many chefs. Simply roast the nuts in a 350°F (175°C) oven for 10 to 15 minutes or until they begin to take on a light brown color, and then run the seeds through the oil press.

Young hickory nuts ripen on the tree.

The seedcake remaining after oil extraction can be utilized as a supplemental feed for livestock, but considering the labor and time involved in the production of hickory oil, the flavorful and edible by-product from oil extraction is much better utilized in the kitchen. It can be enjoyed in baked goods, added to granola and for numerous other culinary applications.

Storage

When kept at temperatures under 45°F (7°C) hickory oil will resist rancidity for up to a year. It's best to store the oil in refrigeration even before opening to prolong its shelf life. Unrefined oils such as hickory and pecan may have the tendency to become cloudy or even partially solidify when stored at low temperatures, but this is to be expected and should not be a cause for concern. The oil will quickly return to its liquid state once back at room temperature.

Uses

Hickory nut oils pressed from raw seeds have a pale yellow color and a light nutty flavor, but roasting the nuts before extraction brings out a bold, buttery flavor that can be enjoyed in a wide range of dishes. Not only can hickory oils be used in salad dressings and marinades, the oil is an excellent choice for frying, baking and sautéing. Heating the oil enhances the nutty aroma and flavor, and all hickory oils, whether pressed raw or roasted, have a smoke point of 470°F (243°C).

While hickory oils are primarily utilized for culinary purposes, there are a number of topical applications in which *Carya* oils are considered beneficial. In particular, the oils are high in vitamins E and A and can be used in lotions or even lip balms for their moisturizing effect. These vitamins are also antioxidants that may help prevent signs of premature skin aging. Hickory oil also contains notable amounts of an amino acid known as L-arginine, which is thought to stimulate hair growth. Also rich in magnesium, pecan oil is sometimes used for its anti-inflammatory benefit in formulas designed to ease the symptoms of arthritis or other inflammatory ailments.

Nigella

(*Nigella sativa*)

The strongly flavored black seeds of *Nigella sativa* have been used as a spice and for their fragrant, dark oil for at least 3,000 years. Native to eastern Europe and western Asia, nigella quickly naturalized throughout Europe and into northern Africa. Remnants of seeds have been found in numerous sites in ancient Egypt, including the tomb of King Tutankhamun, leading archeologists to believe that this area is likely the site of the plant's first domesticated cultivation.

Widely used in Middle Eastern and Indian cuisine, nigella seeds are often referred to as black seed, black cumin, black caraway and kalanji. It's important to note that nigella is not related to cumin or caraway, both of which are from the plant family *Apiaceae*, which includes carrots and parsley, while nigella is from the buttercup family, *Ranunculaceae*. It's also worth mentioning that another species, *Elwendia persica* (*Bunium persicum*), is also commonly referred to as black cumin or black caraway. The seeds of *E. persica* are similarly used as a spice throughout the Middle East, so care must be taken when sourcing seeds to ensure that the proper species is acquired, whether their purpose is for eating, growing or pressing for oil.

Nigella seed oil has been revered for its health benefits among many cultures throughout history. It has been prescribed in ancient Chinese medicine for respiratory issues and improving memory, while the ancient Greeks used black seed oil to ease digestive issues. Traditional Ayurvedic medicinal texts suggest using the oil as a topical treatment for numerous skin conditions, including eczema. Perhaps the boldest claim regarding the benefits of black seed oil was made by the Prophet Muhammad, who said that the oil is "the medicine to heal every disease but death."

The oil pressed from nigella seed is high in omega-6 and omega-9 fatty acids, B vitamins, and many minerals, including iron and calcium. The oil also contains thymoquinone, a compound believed by many to hold promising pharmacological properties against several diseases. The oil is typically ingested as a supplement, offered in capsules and often sold by the bottle in many health food stores. Most commercial black seed oil is cold pressed as high heat is very detrimental to the oil's quality and nutritional value.

The annual flowering garden plant love-in-a-mist or *Nigella damascena*, is also native to parts of Europe, southwest Asia and northern Africa and is sometimes referred to by gardeners as simply nigella. Although these two plants are closely related, *N. damascena* is typically enjoyed for its ornamental beauty while *N. sativa* is utilized for its medicinal oil. Interestingly, the seeds of love-in-a-mist are also edible, with a flavor reminiscent of nutmeg. Perhaps the industrious oil presser could consider running the seeds of *N. damascena* through their machine to trial the qualities of this particular oil? Although these seeds do not contain the compound thymoquinone and therefore may not offer the same health benefits as the seeds of their cousin plant, they may prove to be useful or enjoyable in other applications.

Planting and Growing

Being a small-seeded crop, nigella seeds should be planted at a depth of only ⅛ inch (0.3 cm), and kept moist until germination occurs. Seeds can be direct sown in a full sun to partial shade area after risk of frost has passed, or after the first rains in warmer, long-season areas. Seed crops will take 140–160 days to mature, so many growers may need to start plants indoors in advance of the season. The seeds will germinate faster in warmer temperatures, with the optimum temperature of 70°F (20°C) sprouting seeds in approximately two weeks' time.

Plants started indoors can be set out after risk of frost at a spacing of around 9–11 inches (23–28 cm) between plants and between rows. Wider row spacing can be used if you plan to cultivate mechanically, but nigella seems to perform best when planted close. *N. sativa* is somewhat drought tolerant, but dryness can certainly affect final yields as the plants produce significantly more blooms when kept moderately moist throughout the heat of summer. Once the flowering period has peaked, it's recommended to cease irrigation to allow the plants to begin drying down after reaching maturity.

Plants will produce beautifully intricate flowers over a period of two to three weeks before yellowing and dying back as seed pods develop and mature. If left for too long, the pods will shatter, scattering the seeds, so care must be taken to begin the harvest before this happens. Bringing in the harvest before the pods have reached total maturity may help the grower avoid this potential loss of seeds. Nigella pods can easily be harvested by hand: simply snip them free from the plant and collect them into a bucket or other container. These can then be threshed and winnowed, or screened, to separate the chaff and prepare the seeds for pressing. Alternatively, since the dried pods can be easily shattered, the stems can be cut by the handful and then beaten on the side of the bucket in the field, thus separating and collecting the seeds as you harvest. These seeds will still need to be winnowed or sifted through screens to remove any debris before running them through the oil press.

Large-scale black seed operations make use of machinery to harvest and combine the seeds. One acre of land will produce anywhere from 600–1,000 pounds (270–455 kg), depending on the cultivar being grown, weather conditions and soil quality. Nigella performs best in loose, fertile soils.

Post-Harvest Processing

Once nigella seeds are brought in from the field, they need to be cleaned of any plant material or other debris before being pressed for their oil. This can be accomplished by winnowing the seeds with a fan or by sifting them through a series of screens. First, use a screen with a gauge just big enough to allow the seeds to pass through while holding back the larger bit of pods, dried flowers and leaves, then use a fine-mesh screen that will retain the seeds while the dirt and other fine particles are sifted away.

Allow the seeds seven to ten days to finish drying before attempting to run them through the press. Too much moisture within the seeds will negatively affect yields. Nigella seeds and their oil are temperature sensitive. Heat will degrade their quality as well as their shelf life, so pressing seeds at room temperature is recommended.

Pressing

A basic, turnscrew-style expeller press is the best machine for extracting oil from *N. sativa* seeds. Although the seeds contain anywhere from 30%–40% oil, standard cold-pressing of the seeds will produce only an average of 26% oil, while solvent-based methods produce notably higher, around 37%.

Keep the hopper well fed and, if pressing manually, maintain a consistent turnscrew speed for optimum yields. The seedcake produced by the extraction process can be further ground into a powder for any application, culinary or otherwise, that whole-ground nigella seed might be used for. With an approximate protein content of around 20%, nigella seedcake may also be useful as a supplemental animal feed, if its strong and slightly bitter flavor doesn't bother the livestock.

Storage

Like many other oils, nigella oil should be stored, tightly sealed, in a dark glass or other opaque food-grade container whenever possible. Light, heat and air will quickly cause the oil to become rancid and unusable. Although opinions on storage requirements vary among commercial suppliers, it's generally recommended to store nigella oil in a refrigerator. Temperatures below 45°F (7°C) are ideal for extending the shelf life of this product, and the average refrigerator operates at between 35°–45°F (2°–7°C), making this an ideal storage option. It is best to keep the oil cool as it settles, as well as after decanting, filtering and bottling. Once opened for use, nigella oil will keep for 24 months or longer if kept tightly closed and away from light and heat.

Uses

Nigella sativa oil is most widely utilized as a nutritional supplement, although it can also be enjoyed in various culinary applications and has a long history of use as a topical wellness treatment. While the dose of black seed oil recommended for consumption when used as a supplement varies from one third of a teaspoon up to two teaspoons a day, depending on the anticipated benefit or potential treatment, it's important to consult your physician before beginning a regular nigella regimen. Taken this way, nigella oil can be beneficial for lowering cholesterol levels, easing asthma conditions and lowering blood sugar levels, but it is possible that black seed oil will interact with some prescription medications, and a small percentage of the population has shown mild allergic reactions after regular consumption of the oil.

Since the oil is quite sensitive to heat, it should never be used for cooking, sautéing, or frying but does make an excellent salad dressing oil. It can also be drizzled on finished dishes, such as stir-fries or curries to add flavor and depth to the meal.

Black seed oil is renowned as a topical treatment for acne, eczema and psoriasis, and studies have shown that the oil may also be beneficial in the treatment of skin cancer. The oil can be applied directly to the skin or crafted into a balm or lotion, perhaps including other useful herbs, to improve ease of use as a topical treatment. The ancient Egyptian queen Cleopatra is said to have used nigella seed oil in her hair; in fact, the legendary beauty of her hair is said to have been attained through the use of the oil. The popularity of black seed oil as a scalp and hair follicle treatment still persists to this day.

Peanut

(Arachis hypogaea)

Native to South America, peanuts have become a staple food and oil-seed crop worldwide. The earliest archeological evidence of possible peanut cultivation dates back to more than 7,000 years ago, although it is possible these remains are of a wild species early in the domestication process. Depictions of peanuts can be found in the artwork of pre-Columbian cultures, and widespread cultivation of the crop was well established before Spanish intervention. Spaniards took peanuts with them back to Europe, where they quickly circulated and are now a common crop in western Africa as well as throughout Asia.

Peanuts were brought to North America by enslaved Africans in the mid to late 1600s but remained a small-scale garden crop until the 1930s. At this time, George Washington Carver, an agricultural scientist who helped establish the practice of utilizing nitrogen-fixing legumes such as peanuts and soy in crop rotation to nourish the soil, developed hundreds of uses for the peanut crop, thus launching A. *hypogaea* as a commercial mainstay and altering the landscape of southern agriculture forever.

There are four distinct sub-species of peanut grown in the United States, each utilized for its own particular traits. These four groups are commonly referred to as Valencia, Virginia, Runner and Spanish. It's the Spanish type that is of most interest to the small-scale oil producer as these seeds provide the highest yields, at around 50% oil. These nuts are smaller in size than their counterparts and have a reddish-brown skin. Of course, all peanut types can be pressed for their oil, and may offer slightly differing flavor profiles, but its higher yields make the Spanish type most attractive for cold-pressed production. In the early 1970s a cultivar of runner peanut was developed, known as the florunner, which offered larger harvests and greater disease resistance that what was being grown commercially at the time. This new variety now accounts for more than 85% of total U.S. peanut production, including a majority of what is grown for oil production. Most commercial peanut oil is solvent extracted and heavily refined, with very little color or flavor. This highly processed product is even considered safe for individuals with peanut allergies.

Cold-pressed peanut oil has a deep golden color and nutty aroma and, like many seed and nut oils, is high in Vitamin E. It's commercially available in some locations under the names arachis oil and groundnut oil, although this latter name may cause confusion as groundnut is also the common name of the Bambara groundnut, *Vigna subterranean*, a related species native to Africa. With the ever-growing demand for healthier food options, expeller-pressed peanut oils are gaining a larger share of the commercial market and are a very viable choice for homestead or small business producers.

Planting and Growing

Peanuts grow best in a long, warm season and will need anywhere from 120–150 days to mature. They prefer sandy soil, and need full sun for the best production. Peanuts for seed can be stored in their protective shells but will require shelling before being planted. Plant the seeds one to two inches (2.5–5 cm) deep and about six inches (15 cm) apart. Space rows two feet (0.6 m) apart or three feet (1 m) for runner types.

Around six to eight weeks after planting, the peanuts will begin to flower. The delicate yellow blossoms will appear near the bottom of the plant and, as they are pollinated, will send pegs down into the soil to begin pod development. The specific epithet, *hypogaea*, means "under the earth" in reference to how the plant forms its seeds underground. During this time, it's crucial to provide regular watering, as exposure to extended dry periods during flowering will negatively affect yields.

As the peanuts mature, the leaves and stems will begin to yellow, signaling that harvest time has arrived. Mechanical harvesting is a two-step process that involves first driving a digger over the field to extract the plants. This machine pulls the plants from the ground and flips them upside down, exposing the peanuts to the air. After the peanuts sit in the field a few days to dry, a shaker is driven over the crop to remove the seeds from the plants. This technique can be simulated on a small scale by simply using a garden spade or fork to loosen the soil and then remove the plants from the ground. Be sure that the ground is dry and the upcoming weather forecast looks sunny and dry as well. The peanuts can be left in the field to cure, or if rain or frost is a concern, the plants can be moved to a sheltered area to finish drying. After two to three weeks, once the hulls are completely dry, the peanuts can easily be removed from the plants.

Post-Harvest Processing

After the peanuts have been pulled from their plants, they can be laid out in a critter-free area for another week or so to finish drying. Before being run through the oil press, the seeds will need to be removed from their shells. Shelling seeds by hand can be a tedious process, but considering the peanut's high yield of oil, hand-shelling is reasonable if the goal is simply a small batch of final product. For anything more than this, alternative shelling techniques may be necessary. Replicating commercial shellers is an option and can be accomplished by lightly crushing the peanuts shells with a rolling pin or similar tool, or even running the peanuts through a food processor. On a larger scale, the peanuts can be fed into a small chipper/shredder, which will quickly remove the seeds from their shells. The remaining debris can be winnowed, or the shell-peanut mixture can be added to cold water. The lighter shell particles will float while the peanuts sink to the bottom. With this technique, the peanuts would need to be dried once again before pressing, either in the open air or in a low temperature oven.

Dry roasting the peanuts before running them through the press will give the oil a deep, nutty flavor and dark, golden-brown color. Oil made in this fashion is typically most often used as a flavoring oil as opposed to being used as a cooking oil, similar to how toasted sesame oils are enjoyed, drizzled on meals or as a dressing and marinade.

Pressing

Toasted or raw, peanuts will run through the press better if broken into pieces before being loaded into the hopper. Unless they've been shelled by hand, the seeds have likely been well-crushed during the shelling process. Peanuts will give noticeably higher yields if pressed when warm. In fact, peanuts roasted to temperatures of 320°F (160°C) before pressing have higher oxidative stability and longer shelf life than their raw seed counterparts.

The seedcake resulting from the extraction process is a delicious and protein-rich food that can be preserved for human consumption or utilized as a supplemental feed for livestock.

Storage

Peanut oil is quite stable and can be stored, unopened, in the pantry for up to two years before rancidity becomes a concern. Once opened, the oil should be used within a year. Roasted peanut oils are considered more shelf stable than oils pressed from raw seeds, but both products should be kept in a cool, dark location. Refrigerating peanut oil is certainly possible but will extend the shelf life of the product only an additional two to three months.

Because of its high smoke point, peanut oil is often used to deep fry foods. After cooking, this oil can be kept and reused. Once cooled, the oil should be well filtered to remove any bits of food or other debris, and can then be stored until the next use. Avoid mixing this used oil with fresh, unused peanut oil as this will significantly shorten the shelf life of the oil. Used peanut oil will keep for only two to four months and should be used no more than five times before being discarded.

Uses

Unrefined, cold-pressed peanut oil has a smoke point around 320°F (160°C), but the oil's tolerance for heat can be increased by up to 10 degrees simply by decanting and filtering out any sediment or seed particles. This oil can be used to sauté or stir fry but could also make an interesting salad dressing or marinade. Oil pressed from roasted peanuts can be heated to temperatures around 441°F (227°C), making this product ideal for deep frying or other high-heat applications.

Topical use of peanut oil is less common, although it can occasionally be found listed as an ingredient in commercial products designed to relieve arthritis and joint pain. The oil is high in vitamin E and its emollient properties make it an ideal choice for use in moisturizing, skin-softening beauty products. Its also thought to sooth eczema and protect the skin from UV damage.

Peanut oil has a wide range of industrial uses, including as lubricants, hydraulic fluids and in the manufacturing of margarine. Its greatest potential use outside of the kitchen may be as a biodiesel fuel. In fact, the first diesel engine, invented by Rudolph Diesel and introduced in 1900, actually ran on peanut oil. Rudolph's hope was that a plant-oil–fueled engine could someday become just as well known, and perhaps even replace, the kerosene- and coal-driven machines of his time.

Peanuts should be removed from their shells before pressing.

Poppy

(Papaver somniferum)

The breadseed poppy, more commonly known as the opium poppy, has a long history of use as a medicinal herb, as a food and as the source of the potent narcotic, opium. The use of *P. somniferum* predates written history yet images of opium poppies have been found among ancient Sumerian artifacts, leading scholars to believe that the plant is likely native to the eastern Mediterranean region. Poppies have been used medicinally for thousands of years, a practice that still continues to this day. A variety of opiates are derived from the latex extracted from the green seed pods, including morphine, codeine and papaverine, with

Tasmania, Indian and Turkey being the largest producers of poppies grown for medicinal purposes.

Interestingly enough, the same species of poppy utilized for the production of opium is the source of the edible poppy seeds so popular in baked goods and used for the extraction of poppy seed oil. Due to the plant's potential uses as a narcotic, some countries have heavily regulated the growing, harvesting or possession of opium poppies. In some places, special licenses are required in order to grow poppies. A number of cultivars have been developed as low-morphine selections, but without laboratory testing, it is likely quite difficult to differentiate these newer varieties from the traditional poppies through simple observation in the field. It's highly recommended that one researches the local laws and potential prohibitions involved in growing a poppy crop before getting started.

The specific epithet *somniferum* means "sleep bringing," which refers to the sedative nature of the opiates contained within the seed pods, but the medicines extracted from poppies have also been used by herbalists to treat stomach upset and digestive issues and as an anti-spasmodic and expectorant.

The seeds themselves, and even more so the oil extracted from them, contain very low opiate levels and are generally considered safe for consumption. The oil is high in vitamin E and therefore displays an excellent shelf life. The primary commercial use for poppy seed oil is industrial: it can be found in many paints, varnishes and soaps. In most circumstances, these industrial applications use chemically extracted oils, but cold-pressed poppy seed oil would be an excellent component in small-scale, handcrafted soap production.

Expeller-pressed poppy seed oil is a pale yellow color, while the oil extracted through high-heat or chemical treatment tends to have a more reddish hue. The oil is popular in many skin care products where it is found to be useful as a moisturizer, and its high polyunsaturated fat content makes the oil useful as a rejuvenating hair and scalp treatment.

Planting and Growing
Poppies are easy to grow, preferring fertile, well-drained soil. They will thrive in a full-sun area but still produce a good-sized seed crop when

grown in partial shade. Seeds can be sown in late fall or early spring to ensure exposure to the cold temperatures needed to break their dormancy. Alternatively, seeds can be refrigerated for one to two weeks before planting to achieve similar results. As the seeds are quite miniscule, they should be shallowly planted, perhaps only ⅛ inch (3 mm) deep. Poppies do well when planted in groups, in rows approximately one foot apart. Once seedlings emerge, they can be thinned to around six inches (15 cm), but if planting a large area, it may be easiest to simply plant the seeds at this distance apart. A large field of poppies can be planted by simply broadcasting the seed onto the soil, but adequate moisture levels must be maintained until the seedlings are well established.

Young poppies often struggle when competing with weeds, so a thick mulch is recommended if possible. This will also help retain moisture in the soil. Keep the crop evenly irrigated, especially during the peak heat of summer, until flowering begins, and then cut back on watering. After the flowers have died back, the green seed pods will continue to develop and mature. Once the pods begin to dry down, they will turn grayish-brown in color, and this is when the seed harvest will begin.

Post-Harvest Processing

Many breadseed poppy varieties will have open vents along the top of the mature seed pods, meant to aid in dispersal of the seeds, but some cultivars have been developed with sealed vents to help guard against seed loss in the field. If the poppies being grown are of the more traditional, open-vent types it's important to monitor the crop and harvest as soon as the seeds can be heard rattling within the pods. Waiting too long will increase the chances of losing seeds as the top-heavy plants tend to lodge, or fall over, once the stems have begun to dry and become more brittle.

Harvesting poppy seeds is relatively straightforward regardless of the size of the plot. Stems can be cut and gathered by the bunch, upside down, in buckets, baskets or other containers. Be sure to use containers that will be able to catch and hold the seeds that drop from the pods, avoiding any open weave-type baskets that may allow seeds to pass through. Each seed pod will contain approximately 200 seeds, producing an average of 2,500 seeds per square foot, so a usable amount of poppy seeds can be grown and harvested from a relatively small plot.

Agitating the seed pods by simply beating them along the inside of a bucket will be all the effort required to release the seeds, which will just pour free from the dried seed pods. Sealed-vent pods can easily be broken open to release the seeds. The poppy seeds can be further cleaned of any debris by sifting them through a series of screens in preparation for oil pressing.

Pressing

Once poppy seeds are dry enough to release from their pods, they are almost dry enough to press. Allow the seeds to air dry, in front of fans, for up to a week before pressing. Be sure to protect the seeds from rodents. Dried poppy seeds can be stored in airtight, food-grade containers, held at temperatures from 45°–60°F (7°–15°C) until pressing time, if needed. The poppy seeds can be expeller pressed at room temperature and will yield anywhere from 40%–50% oil, depending on variety and growing conditions.

The resulting seedcake is edible and retains the nutty flavor of the seeds. It can certainly be used to feed livestock if desired but can also

Delicate poppy flowers can add texture and color to a garden bed.

be utilized in culinary applications for human consumption as well. Traditional Hungarian recipes call for ground poppy seeds, which are often boiled in milk and sugar with other ingredients and then used as a traditional filling in a number of dishes, including palacinta, kifli and strudel. The poppy seedcake left over from the oil extraction process can be enjoyed in much the same way.

Storage

Due to its high vitamin E content, poppy seed oil is considered quite shelf stable and an unopened container will keep at room temperature for 18 months or longer. Stored at 40°F (4°C), poppy seed oil will resist rancidity for three to four years. Properly decanting and filtering the oil will further improve its storability. Seed particles left within the oil will promote rancidity and negatively affect the quality of the product.

Uses

The smoke point of a refined, commercial poppy seed oil is around 400°F (205°C) but expeller-pressed poppy seed oil should not be used for high-heat cooking. In fact, this oil is considered one of the thermally stressed culinary oils, which can quickly go rancid when exposed to high temperatures. It's recommended that poppy seed oil be used only for low-temperature culinary applications such as salad dressings and marinades and as a finishing oil on meals.

Poppy seed oil contains high amounts of linoleic acid, a polyunsaturated omega-6 essential fatty acid. Because of this, the oil is quickly absorbed into the skin and is known to be moisturizing while protecting skin and hair from damage. The oil has a relatively light odor and texture, which makes it an ideal choice for use in skin care products as well as handcrafted soaps. The light color and scent of poppy seed oil is ideal for herbal extractions used for crafting lotions and balms and as a carrier oil for formulations involving essential oils.

Pumpkin (*Cucurbita* spp.)

The typical orange field pumpkin is one of many cultivars of the species *Cucurbita pepo*. This species includes most summer squashes, some gourds as well as a few of the winter-type squashes, which, as their name implies, are most commonly stored for use in the winter. The cucurbit family also includes other species of domesticated squash, with *C. maxima* and *C. moshata* being the most common. The seeds from any of these species are considered edible and can most certainly be pressed for their oil, although some particular varietals of *C. pepo* have been developed specifically for use as oilseed crops.

These various species of squash are native to Central and South America, and archeological evidence discovered in the Oaxaca Highlands of Mexico suggests that *C. pepo* was first domesticated more than 5,000 years ago. These fruits quickly became a staple of the Indigenous diet. Many varieties were enjoyed for their edible flesh, which could be dried and eaten throughout the winter, but the seeds, with their high protein and fat content, were considered the most valuable part of the harvest.

Columbus's writings reveal that he first viewed a cultivated pumpkin field on the island now known as Cuba in December of 1492, and it's commonly believed that he took squash back to Spain after his first expedition. The earliest written reference to pumpkins in Europe can be found in the prayerbook of Anne de Bretagne, the Duchess of Brittany, from the early 1500s. From Spain, pumpkins quickly spread throughout the continent and beyond, although various species of gourds, such as the bottle gourd of Africa (*Lagenaria siceraria*) and the wax gourd of Asia (*Benincasa hispida*) were already somewhat familiar at the time.

It was in central and eastern Europe that the seeds of various pumpkins were first pressed for their oils, as documented in an Austrian reference book from 1739. Just 34 years later, in 1773, the Austrian Empress Maria Theresa enacted a law stating that "this healthy pumpkin seed oil is unique and much too precious.... [I]t shall not be used as a culinary delicacy anymore but shall be collected and distributed only by the apothecaries." However, over the next century, pumpkin seed oil production increased at such a rate that this mandate was lifted as the supply far outweighed the demand required by pharmacists crafting medicines, balms and ointments from the oil.

In 1870, a Styrian pumpkin farmer in southeast Austria discovered a random genetic mutation in his fields, which resulted in a crop of *C. pepo* that produced seeds with a greatly reduced testa, or seed coat. These essentially hulless seeds proved ideal for oil extraction, and the resulting oil was considered superior in flavor and quality. Over time this "naked" seed trait was stabilized, and the Austrian pumpkin seed oil market was born!

Since that fateful summer day, a number of hulless-seeded pumpkin varieties of *C. pepo* have been developed. These seeds are ideal for roasting, snacking and particularly for oil production. Unfortunately, the flesh of most of these cultivars is quite bland and typically used for animal feed, although breeders are always working to improve the flavor and thus increase the usefulness of these unique pumpkins. Of course, as mentioned earlier, seeds from any domesticated squash can be pressed for oil, and some varieties, such as butternut (*C. moshata*), are prized for their flavorful and nutritious oils.

Planting and Growing

Pumpkins grow best in warm weather and typically take 90–100 days to mature, although this can vary depending upon the variety grown. They prefer direct sowing after any risk of frost has passed and soil temperatures have reached 70°F (21°C), but growers in short-season areas can start their seeds indoors about three to four weeks before planting out, being cautious to prevent the seedlings from becoming rootbound. The plants should be given plenty of space to sprawl, as most varieties of pumpkin produce aggressive vines.

Regardless of the variety of pumpkin or squash being grown, the fruits will signal their maturity by changing color as they ripen; the rind will be firm, and the plants will begin dying back. The technique for harvesting the ripened fruits and extracting their seeds will depend upon the size of your garden as well as your needs as a grower. Collecting a significant quantity of pumpkin seeds for oil production can be a laborious task without the use of specialized equipment.

Large-scale farms make use of a tractor-pulled machine that positions the pumpkins in the field in a straight line; they are then picked up with a large, spiked wheel and dropped into extraction drums that crush the fruits and separate the seeds. A smaller version of this machine, pulled on a trailer, is a viable option for mid-scale growers producing pumpkin seed oil commercially or cooperative operations pooling their resources. With these machines, pumpkins are loaded into the pulverizer by hand and then fall into a rotating, screened drum below, where the seeds are separated from the mash. At this point, the seeds should be collected into buckets or barrels, water added and the mixture left out to ferment for two to three days.

This fermentation process will help break down any remaining flesh and aid in the final cleaning process. The seeds should then be washed, using running water and screens to help remove any fruit or plant debris, and then laid out in front of fans to dry.

Post-Harvest Processing

Until the chance discovery of the hulless-seeded pumpkin variety, all squash seed oils had been pressed from whole seeds, with the shell intact. While it's certainly possible to shell seeds before pressing, this may prove too laborious a task to be considered worth the effort. Pumpkin seeds can be boiled or roasted to help loosen the shells and then pounded or milled to separate the seeds from their shells. The broken seeds can either be winnowed to remove the chaff or added to cold water, and the shells, which will float, can be skimmed off the surface.

This arduous task can easily be skipped, either by pressing the seeds whole, or by choosing one of the many hulless-seeded varieties now available. These naked seeds are said to produce the finest oils, as pressing the shells tends to add a slightly bitter flavor to the final product. Additionally, the hulless varieties generally contain a higher oil content than the in-shell cultivars.

Regardless of which variety is being pressed, it's recommended that the seeds are first roasted, or at the very least heated, before pressing. This will increase yields and improve flavors.

Pressing

After heating, the pumpkin seeds can be first crushed and then pressed or simply fed into the expeller press whole. Lightly crushing the seeds before pressing will typically increase yields, but trialing both methods will help you determine if the additional step is advantageous for your

Roasted pumpkin seeds
ready for the press.

situation. The resulting oil will need to settle for anywhere from one to three weeks, with oils pressed from in-shell seeds needing the longest time to clarify. At this point, the oil can be passed through a filter if desired and bottled for storage.

Storage

Pumpkin seed oil is best stored in dark glass with a tight-fitting lid, as both sunlight and air will expedite rancidity. Oils pressed from hulless-seeded varieties tend to be higher in vitamin E than their hulled counterparts. Vitamin E works as a preservative, and these oils can be stored at room temperature unopened for up to 12 months. The ideal temperature for storage is 64°F (18°C), but once opened, the oil should be used within six to twelve weeks. Other squash seed oils can be refrigerated to prolong their shelf life and will keep this way for up to a year.

Uses

Pressing hulless pumpkin seeds produces a beautiful oil, with colors from light and dark green to a deep, dark red, depending on the thickness of the sample. This is due to an optical phenomenon known as dichromatism. Oils extracted from other squash, such as butternut, tend towards a golden, amber color.

These oils are considered some of the healthiest culinary oils available, and should be used only in low- to no-heat culinary applications. The smoke point of the oil is around 248°F (120°C); therefore pumpkin seed oil is not suitable for frying or most cooking but is reserved for salad dressings, marinades and similar applications.

Pumpkin seed oil has a long history of use as a medicinal supplement, touted for its many benefits, including lowering cholesterol levels, improving hair growth, serving as an anti-oxidant and even boosting the immune system due to the oil's high levels of zinc and iron. The oil is also thought to be quite beneficial for topical use and is included in many commercial skin care products to moisten and firm the skin as well as to soothe inflammation, treat acne and ease the redness and irritation caused by eczema. Pumpkin seed oil is so highly prized by cosmetics manufacturers that it is often referred to as "green gold."

Safflower

(*Carthamus tinctorius*)

Native to southern Asia and west into Africa, safflower has been utilized as an edible flower, as an oilseed crop, and for producing a strong yellow dye since the days of ancient Egypt. Remnants of safflower plants have been found in the tomb of King Tutankhamen, and archeological evidence dating back to 2500 BCE suggests Mesopotamia as the likely location of safflower's original domestication. The specific epithet *tinctorius* means "of or pertaining to dyeing" and refers to the plant's history of use in producing a dye for clothing and other fabrics. In some parts of the world, safflower petals have been used as an inexpensive substitute for saffron to add color to rice dishes, and it is widely thought that the common name, safflower, is derived from this particular use.

With the advent of synthetic dyes, the use of safflower for this purpose is rarely seen outside of Asia or among small-scale artisan crafters. The most common modern use for safflower is as an oilseed crop. India is the largest commercial producer, followed by the United States, where California is the center of commercial production. Safflower will easily reseed itself and has become naturalized in many parts of the western United States. The seeds are also commonly included in commercial bird seed mixes, and safflower plants can sometimes be found in areas near bird feeders or in the wild where birds may have distributed seeds collected from a residential bird feeder.

Most safflower oil is produced through chemical extraction methods and utilized for a number of industrial purposes, including use in paints and varnishes, although a large amount of safflower oil is dedicated to the manufacture of margarine. This refined oil product can also be found in many commercial salad dressings and cooking oil blends, but expeller-pressed alternatives are becoming more common thanks to consumer awareness and increased interest in healthier food options.

Expeller-pressed safflower oil is a delicate, straw-yellow color and has a very light scent. The oil is rich in unsaturated fats and is notably high in vitamin E. Because of this, safflower oil is considered a healthy choice for high-heat culinary applications. Additionally, due to the oil's high vitamin E content, its also thought to be beneficial for dry and irritated skin and is employed for this purpose in numerous health and wellness products.

The oil content of the seeds will range from 30%–40%, depending on the cultivar and growing conditions. Most safflower plants have spines along the leaves and bracts, surely a defense mechanism to protect the seeds from predation, so care should be taken when harvesting seeds by hand. Recent breeding work has spurred the development of spineless safflower cultivars, but these newer types tend to yield less oil than the traditional varieties.

Planting and Growing

Safflower plants perform best in fertile, deep soil and require hot, dry weather for the greatest production and highest oil content. The long taproot allows the safflower plant to be quite drought tolerant, and, in fact, overly wet soils will stunt the plant's growth.

Seeds should be planted approximately half an inch deep in rows 6–12 inches (15–30 cm) apart. Keep the seedbed evenly moist and expect up to two weeks until germination. Within the first two to three weeks of growth, safflower will form a dense rosette of leaves close to the ground while it establishes its main taproot. At this stage, the plant is tolerant to temperatures as cold as 20°F (–6°C), but once the stems begin to form, safflower is significantly more susceptible to frost damage. Plan for 110–120 growing days until seed harvest. As leaves begin to turn yellow and die back, around 40–45 days after flowering, check the dried flower heads to assess maturity. The seeds should easily come loose from the dried flower heads when rubbed between the hands. Once the crop is ready to harvest, do not delay, as waiting too long will result in potential seed loss from shattering as well as from birds. Small crops can be protected from birds by netting, but larger fields are certainly more vulnerable to predation.

Small plots of safflower can easily be harvested by hand. Simply snip the bunches of flower heads and collect them in baskets or buckets. Wearing sturdy leather gloves while harvesting will protect your hands from the spiny leaves and bracts. Once collected, the flower heads will need to be threshed in order to release the seeds. This can easily be done on a small scale by hand or with the use of a flail or other similar tool. The threshed seeds will then need to be winnowed or sifted through a series of screens to remove any chaff. Pouring the seeds from one container to the next in front of a fan is a quick and effective method for removing any debris from the seed harvest. Larger operations can make use of a small-grain combine to harvest and clean their safflower seed.

Post-Harvest Processing

Safflower seed shells are quite hard, and many commercial operations will hull their seeds before the oil extraction process. This is usually accomplished with a dehulling machine that utilizes a series of rollers to crack the seeds, which are then passed through screens in front of fans to completely remove the shell debris from the seeds. This can be replicated on a smaller scale using a household rolling pin, but this can be a slow and ineffective process. Alternatively, the seeds can be crushed in a food processor and then added to a container of cold water. The shells, which are lighter, will float, while the seeds will sink to the bottom. With this technique, the seeds would need to be thoroughly dried again before running them through the oil press.

Whole safflower seeds can be run through the oil press, but this may prove to be more difficult with a manual, hand-turned machine. A manual machine upgraded with pedal power or a motorized oil press should have no issue crushing the hard shells of the safflower seed. If you're using a hand-turned machine, consider milling the seeds in a food processor, thus breaking open the hard shells, to alleviate some of the work.

Pressing

Pressing safflower seeds for oil is a straightforward process once the seeds have been cleaned and allowed to dry for at least one week after harvest. Seeds can be stored in food-grade buckets or other containers, to protect them from moisture and rodents, until pressing time.

The seeds can be pressed at room temperature, whether whole or milled, but consider gently heating the seeds before pressing, as this may increase oil yields by up to ten percent. To maintain maximum nutritional value and optimum flavor, do not heat the seeds to temperatures greater than 120°F (49°C).

The seedcake resulting from the extraction process is a valuable source of protein and is an excellent supplement to livestock and poultry feeds. If the seeds were pressed whole or milled along with their shells, the seedcake is not recommended for human consumption. If hulled seeds were used in the oil extraction process, the resulting seedcake can be used in baked goods, blended with cereals or similar culinary applications.

Storage

Since safflower oil is high in polyunsaturated fats, it's ideal to keep the oil well sheltered from heat, light and air. The optimum storage temperature is at or below 57°F (14°C), so an unopened, opaque bottle will keep in a pantry for up to two years. Once opened, if stored at room temperature, the bottle of oil is likely to become rancid within six months. Refrigerating safflower oil will increase its shelf life dramatically. An opened bottle of safflower will resist rancidity at temperatures at or below 40°F (4°C) for one to two years. Refrigeration may cause the oil to become cloudy, but this in no way affects its quality.

Uses

Safflower oil is an excellent choice for high-temperature cooking. Unrefined, cold-pressed oil has a smoke point of 225°F (107°C), and well-filtered, semi-refined oil can be heated to 320°F (160°C) before smoking occurs. This means that safflower oil is ideal for frying, sautéing and baking. The oil is also useful for salad dressings, marinades and other low-temperature culinary applications, where its neutral flavor pairs well with a wide variety of dishes.

Thanks to its high content of linoleic acid, safflower oil is well regarded as a topical moisturizer and can be crafted into any assortment of balms or lotions for this purpose. Commercially available treatments for acne and aging skin often contain safflower oil among their ingredients. Safflower oil-based products are considered anti-inflammatory and are beneficial in relieving rashes and other issues involving irritated skin. With regular topical use, the oil is reputed to enhance the skin's texture, appearance and quality.

Sesame
(*Sesamum indicum*)

An ancient oilseed crop, the domesticated species of sesame was first cultivated in India more than 5,000 years ago. The crop quickly spread, and archaeological evidence suggests that sesame was being grown in Turkey by 700 BCE and in Egypt by 300 BCE.

Sesame is a favored crop in less hospitable environments, due to its robust, rugged nature and drought tolerance. Sesame thrives in high heat with little moisture and has been found to be particularly successful when grown at the edge of deserts where other crops simply cannot thrive. It's interesting to note that sesame is mentioned in the *Papyrus Ebers*, an Egyptian book of medicine written sometime around 1550 BCE, but a wild species was likely discussed in the book, as opposed to the domesticated *S. indicum*, which had not arrived in the region by that time. Most wild species of *Sesamum* originated in Africa, and it may have been any one of these wild forms that was documented in the ancient medical text. Historians and archeologists debated for quite some time whether domestic sesame originated in Africa or further east in India, but recent evidence suggests that the genetic precursor to modern sesame is *S. orientale* var. *malabaricum*, the singular wild species found solely on the Indian subcontinent.

There are a number of varieties of domesticated sesame available to the oil presser, with seed coats diverse in color—white, tan, red, brown, and even black. While many modern varieties, bred to have non-shattering pods for ease of mechanical harvest, have white seed coats, it is commonly believed by many that the oil extracted from brown-seeded sesame results in the most flavorful oil. In fact, most oilseed crops of sesame, grown extensively in India, are produced from brown-seeded types, while white-, black- and red-seeded sesames are typically reserved for culinary applications, such as baking and seasonings, and even medicinal uses.

Traditional benne, the first sesame brought to North America by enslaved Africans, is a brown-seeded varietal that produces a deeply colored, golden oil. (The name benne, which is widely used in the southern U.S. when referring to sesame, comes from the word *bene* in the West African Bambara and Wolof languages, whereas sesame comes from the ancient Egyptian *sesemt*.) This varietal was almost lost when growers moved to larger-scale commercial farming of sesame and the convenience of the newly developed, non-shattering types overshadowed their traditional predecessors. Thankfully, due to the dedication of a handful of historians, preservationists and farmers, the traditional benne has been saved and is available for growers interested in working with this heritage sesame.

Planting and Growing

Depending on the cultivar, sesame can take anywhere from 90–135 days to reach maturity. In many places, the seeds will need to be started indoors around four weeks in advance and transplanted out into the field when temperatures have reached around 70°F (20°C). In areas with cooler spring temperatures, the young plants can be kept covered until mid-May to June, when the night temperatures are warmer. Black plastic can also be laid out in the field before transplanting to increase soil temperatures. The more effort put towards keeping the plants warm, the more productive the sesame. Space plants about six inches (15 cm) apart. Although sesame is a drought-tolerant plant, the seedlings will need moderate watering until they are established.

As the plants reach maturity the seed pods will begin to brown, beginning near the bottom of the stem. Before wet weather arrives, gather your sesame by cutting the plants at the base and laying them out flat in a sheltered area to finish drying. Do not wait too long to harvest or you risk losing seeds in the field. Once the plants have completely dried, the seeds can easily be threshed by beating the plants on the inside wall of a bucket, thus releasing the seeds to be collected below. At this point the seeds will need to be cleaned of debris, either by sifting them through a series of screens or by winnowing the seeds with the help of fans. Sesame seeds are quite light and may prove difficult to winnow, but with some practice, this technique may be the quickest for cleaning larger quantities of seeds. If you're growing a non-shattering variety and are planning to harvest mechanically, you must wait until the entire plant has dried and all of the pods have properly ripened before harvesting.

Post-Harvest Processing

After the sesame seeds are threshed and cleaned, they are ready to be pressed. Store your sesame seeds in tightly sealed, food-grade buckets or similar containers until pressing time to ensure the seeds cannot absorb moisture from the environment. Large-scale, commercial oil producers expose their seeds to a dehulling process before extraction, and, if desired, the home-scale oil presser can dehull their seeds in a relatively similar fashion.

Large-scale dehulling of sesame seeds involves soaking the seeds in a high-temperature lye mixture to loosen the hulls, followed by a run through a mechanical dehuller. The resulting mix of shells and seeds is then submerged in a water tank, where the floating debris is skimmed clear and the seeds, which are heavier, sink. The seeds are then dried with hot air while being sorted for size through a series of screens. This technique can be replicated on a much smaller scale at home but may prove to be significantly more effort than one may want to exert. If the DIY oil presser requires hulled seeds, it may be simpler to find a wholesale supplier to purchase from.

There are two distinct sesame oil products available, the golden-colored oil with its delicate flavor and high smoke point, and the darker, noticeably stronger-flavored oil used as a seasoning, most typically in Asian style foods. While the first is produced by pressing raw sesame seeds, the second, darker oil is a result of first toasting the seeds before running them through the oil press. Toasting can be done on the stovetop, but using the oven will allow for more even heating of larger quantities at one time. Simply spread seeds out, in a single layer, on a baking sheet and place in a 350°F (180°C) oven for approximately five to eight minutes, then quickly transfer the seeds onto another surface to cool. Sesame seeds toast quickly and care must be taken to not burn them.

Pressing

Oil pressed from raw, room-temperature sesame seeds is thought to provide the highest quality oil, but gently heating the seeds to 160°F (70°C) before pressing will produce a moderate increase in yields. Toasted sesame seeds can be set aside to cool or pressed immediately after being removed from the oven, and again, warmer seeds will provide greater yields.

The turbid oil should be set aside, well-labeled, in a cool dark place to settle for two to three weeks before decanting and bottling. Sesame oil also benefits from filtration as the sedimentary particles are quite fine. This additional filtration will improve clarity and increase the shelf life of the oil.

The seedcake resulting from the extraction process can be used in many of the same ways one would use sesame seeds, whole or ground. It can be included in breads or other baked goods, made into tahini, added to smoothies or even incorporated into a batter for fried fish or chicken.

Storage

In comparison to other high-smoke-point cooking oils, sesame oil is the least prone to rancidity. This is due to natural antioxidants, such as sesamol, found in the oil. Stored at room temperature, an unopened bottle of sesame oil will easily keep for a year, and if refrigerated, the oil will stay consumable for two years or more. Once opened, the oil can be kept in a cupboard for a minimum six months before rancidity is a concern and for a year when stored in refrigeration.

Uses

The golden-yellow oil pressed from raw sesame seeds has a light fragrance and delicate flavor. It can certainly be used as a salad dressing oil or drizzled on foods before serving, but with its high smoke point of 410°F (210°C) this oil is a great choice for sauté and stir-fry applications. On the other hand, the umami-flavored oil produced by pressing toasted sesame seeds has a lower smoke point 350°F (175°C) and is typically used to add flavor to dishes after cooking or right before removing them from the heat. The toasted oil is notably more resistant to oxidation and remains shelf-stable longer than raw sesame oil.

Aside from its culinary uses, sesame oil has been utilized as an ingredient in cosmetics, soaps and numerous topical wellness products. Ayurvedic medicine has long recommended sesame oil for use in massage, and in China the oil pressed from black-seeded sesame is touted for its ability to tone and tighten aging skin.

Toasted sesame seeds produce a much darker oil (*left*) than the oil extracted from raw seeds.

Sunflower (*Helianthus annuum*)

Sunflowers are one of the few commercial food crops native to North America. The flowers are believed to have been domesticated for cultivation more than 4,000 years ago. There is some speculation among scholars that Indigenous populations in some areas likely pressed sunflower seeds to extract their oils, but with their edible leaves, sprouts and seeds, the plants were mostly utilized as a food crop.

The Spanish brought sunflowers to Europe in the early 1500s, and they quickly spread throughout the continent and beyond, yet they were mostly thought of as a curiosity and enjoyed for their beauty. Once sunflowers reached Russia, they were quickly adopted, and by

1830, sunflower oil was being produced on a commercial scale. The Russian government funded research and breeding programs that led to the development of two distinct types of sunflower seeds: large-seeded varieties meant for human consumption and smaller, oil-type seeds designed for high-yield oil production. Fifty years later, these improved sunflower varieties made their way back to North America, and by 1930 large-scale sunflower oil production had begun in the U.S. and Canada. The most common oilseed variety was the Russian Peredovik Sunflower, which is still widely grown today.

The small, typically black-shelled, sunflower seeds are ideal for pressing, yielding anywhere from 24%–47% oil. These are the same sunflower seeds that are also commonly sold as bird feed. While there are many hybridized sunflowers on the market today, including those bred specifically for oil production, the most common varieties available produce an oil that is high in polyunsaturated fats, omega-6 fatty acids and vitamin E.

Planting and Growing

Sunflowers are relatively easy to grow, although deer find the tops of the young plants quite enjoyable, and predation can certainly be an issue early in the season. In smaller gardens, this can be avoided with some chicken wire or other fencing, but larger fields may need more than just fencing to protect the crop. Many growers have developed their own unique methods for repelling deer, such as soap, human hair or urine strategically placed around the growing area or interplanting deterrents such as foxglove or poppies or strongly fragrant herbs such as sage or lavender. Commercial deer-repellant sprays are also available, but an electric fence may be the most effective method of deer control for those willing and able to make such an investment.

As their name suggests, sunflowers grow best in a full-sun area and will get rather lanky in the shade. They prefer loose, well-drained soil and should be planted after danger of frost has passed and soil temperatures have reached around 60°F (15°C). Seeds should be sown about one inch (2.5 cm) deep and then thinned to 12 inches (30.5 cm) between plants. Sunflowers can be started in advance of the season and planted out when soil warms, but due to the plant's deep taproot, direct sowing is recommended whenever possible.

Harvesting

Leave your sunflowers in the field to mature until the foliage begins to yellow and the back of the flower heads turn a brownish-yellow color. The seeds will appear plump and loose. The mature flower heads can easily be harvested by hand. Cut the stalks about six inches (15 cm) below the flower and toss the heads into a basket. The seeds can be then be freed from the flower head by simply rubbing them loose by hand or over a hardware cloth or screen, allowing the seeds to fall and collect in a basket, tote or other container. Some innovative seed collectors have utilized the spinning wheel of a bicycle to quickly knock the seeds free from the flower head and onto a tarp.

Post-Harvest Processing

Once the seeds are removed from the plants, they'll need to be cleaned of any dried flower petals, leaves, dirt or other miscellaneous debris before they can be run through the oil press. This can be accomplished by sifting the seeds through a series of screens, or by using a box fan to gently blow the chaff out of the seeds, a process known as winnowing.

The seeds should then be spread out on screens in front of fans to dry. Precaution should be taken to prevent rodents, birds or other pests from snacking on or otherwise contaminating the harvest. While commercial oil manufacturers use low-heat drying chambers to dry the seeds to the optimum moisture level of 5%–10%, the small-scale producer can simply let the seeds dry on the screens for seven to ten days. The seeds are then ready to be pressed, or can be stored in airtight, BPA-free, food-grade plastic buckets until pressing time.

Sunflower seeds can be run through an expeller-type press whole, in shell, but some producers prefer to shell, crush or even heat the seeds before extraction. The seeds can be gently heated in the oven before pressing, which will moderately increase yields, but this can prove cumbersome in a household setting when pressing large quantities of seeds. Maintaining a temperature below 100°F (38°C) is ideal for retaining the highest-quality product.

The seeds can be crushed using a food processor or other similar equipment, increasing the surface area to be pressed and thus improving yields, but again, this task will be time consuming and the increase in yields may not warrant the additional effort. Experiment with different techniques until you find what works best for your individual situation.

While the seeds can be pressed in-shell, the resulting seedcake will then, of course, contain shell and will be unfit for human consumption. This seedcake still has a number of valuable uses, including as a high-protein livestock feed or a nutrient-rich compost. If an edible seedcake is the objective, the sunflower seeds will need to be shelled before pressing.

Larger, commercial-scale operations utilize pneumatic shellers to hull the sunflower seeds, but most oilseed varieties, with their thin shells, are left intact. Seeds that are shelled are then winnowed to remove any debris. At home, seeds can be shelled using a rolling pin or food processor to break open the seeds. The broken seeds can then be placed into a bowl of cold water to separate the shells, which will float, from the seeds, which will sink. If this method is used, the seeds will need to again be dried before running through the oil press.

Pressing

Whether the seeds are whole, shelled or crushed, the pressing process remains the same. Simply load the seeds into the hopper once the press is properly prepped and operating.

If you're pressing whole, in-shell seeds you'll notice that the residue and debris that collects in the turbid oil is thicker and darker than what is produced when pressing shelled seeds. This oil will need to sit a bit longer to settle before being decanted, as the particles from the shell itself are quite small. Filtering the oil is also recommended to improve clarity as well as shelf life and even the smoke point, as the small particles are likely to burn when heated.

Storage

Like any seed or nut oil, sunflower oil is best kept in a tightly sealed container in a cool, dark location. The optimum storage temperature range for this oil is 40°–70°F (5°–20°C), so a cupboard or pantry is ideal. Be sure to properly label your oil with the date of extraction. In proper storage conditions, an unopened container of sunflower oil can keep

Brilliant, golden-colored sunflowers are an easy-to-grow, annual oilseed crop.

for up to two years. If regularly opened, a bottle of sunflower oil may stay fresh for only a year before turning rancid, in which case the oil will take on a bitter or somewhat soapy smell.

Uses

The unrefined oil from sunflower seeds has a rich, golden color and a mild, nutty flavor. In the kitchen, sunflower seed oil can be used in many of the same ways that one would use virgin olive oil, such as salad dressings or marinades. While the refined oil can be used to sauté or fry, the smoke point of expeller-pressed, unrefined sunflower oil is around 225°F, so this is not recommended for high-heat cooking. Sunflower oil is light bodied and rich in vitamins A and E, which makes it valuable for use in skin and haircare products.

Approximately 40% of all sunflower oil produced in North America is converted into biodiesel fuel, and the resulting seedcake is utilized as livestock feed and fertilizer. The process of transforming vegetable oils into biodiesel is known as transesterification. This process can be replicated at home using sunflower oil, methanol, hydroxide and water, but it does require special safety precautions as this chemical process involves caustic chemicals and can be hazardous. Be sure to research thoroughly before trying this at home.

Walnut
(*Juglans* spp.)

When considering walnuts as a potential oilseed crop, determining which species of this edible nut to work with is the first step. Most, if not all, commercially available walnut oil is produced from the domesticated English walnut, *Juglans regia*. This nut is also known as the Persian or Carpathian walnut and is widely considered to be the most common walnut. Thought to originate in the area now known as Iran, *J. regia* was introduced into the western Mediterranean region by the 4th century BCE, when growers worked to develop more lateral branching and larger fruits. English walnuts made their way east into Asia along the silk road, while the improved varieties traveled west with the ancient Romans, and were eventually brought to North America by English colonists. This walnut is now cultivated around the world for its edible nut, although a majority of commercial production occurs in China.

North America is home to a number of native species of *Juglans*, most notably the black walnut, *J. nigra*, as well as the white walnut, *J. cinerea*, which is commonly referred to as the butternut. Although black walnut is enjoyed for its flavorful nut, it is most commonly cultivated in the United States, as well as in Europe, for lumber, which is valued for its deep brown color and workability. Its important to note that all species of walnut produce an edible kernel, but some are less desirable for cultivation and use, having thick shells and relatively small seeds.

English walnuts have been pressed for their oil since the Middle Ages, when it was used as a lamp oil, and Renaissance painters prized the oil as a paint base. The earliest records of walnut oil production for culinary use can be found in the Swiss city of Neuchâtel, an area historically known for its walnut oil products. These 16th-century documents describe two distinct oils, one resulting from cold pressing walnuts, considered the superior oil, and the other made using high heat, leading to higher yields.

Indigenous North Americans utilized the various native *Juglans* for their edible nuts, and with their relatively high oil content, which averages around 60%, simply crushing the nuts would make an oily mash useful for a number of applications. It wasn't until around 1630, when black walnuts were brought into Europe, that the seeds were first pressed for their oil using a mechanical extractor similar to a ram press, although greater yields can certainly be accomplished with a turnscrew-style machine.

Planting, Growing and Foraging

Growing walnuts from seed can be a long process (about ten years from seed until nut harvest) but certainly possible with a little patience. Freshly harvested and hulled walnuts should first be placed in water to test their viability. Healthy, mature nuts will sink in the water, while walnuts that are not viable tend to float. The viable seeds will then need to be exposed to a period of cold temperatures, a process known as stratification, in order to germinate. This can easily be accomplished by placing the nuts, still inside the shell, into a moist medium such as sand and peat moss or even potting soil and then storing them in a cool location for three to four months. A refrigerator is the perfect place to stratify walnut seeds. The walnut shells should then be gently cracked, and planted about one to two inches (2.5–5 cm) deep in the soil. While the seeds can be planted directly into the ground, some growers prefer to start their seedlings in containers and then move the trees out into the orchard once the plants are more established. Either way, be sure to keep the seeds, as well as the young seedlings, well watered.

Alternatively, walnut trees can be purchased from a local nursery and planted directly into the orchard. The trees prefer full sun and will tolerate most soil types. Walnuts are only partially self-fertile, so planting multiple trees will increase yields. Expect to wait up to 10 years before trees mature and produce walnuts. Grafted trees purchased from a nursery will fruit sooner and may even produce a crop within five years of purchasing. It's important to note that walnut trees exude juglone, a chemical toxic to many other plants. While this is the tree's method of suppressing any competition, it can also prove quite frustrating for gardeners as juglone negatively affects many garden crops, including tomatoes, peppers, eggplants and potatoes. Other crops are more tolerant and should be considered if growing near walnut trees. Some options include beans, melons, squash and carrots. English walnuts exude notably less juglone than black walnut trees and therefore may not present the same challenge to gardeners.

While stewarding a young walnut tree from seed to harvest is certainly a noble endeavor, some might find that the abundant black walnut trees growing wild across North America provide plenty of walnuts for their oil needs. Foraging wild-grown walnuts is an excellent alternative to cultivating a walnut orchard, although these wild species can be more difficult to process than the domesticated English walnuts. The hulls of *J. regia* will actually split open upon maturity, unlike any of the wild species, and therefore are much easier to remove.

This splitting of hulls is the English walnut's signal that the nuts are mature and ready to be harvested. They can be picked by hand from the lower branches, while the nuts from higher in the tree can be knocked loose by a long pole, perhaps with a hook on the end used to grasp and shake the branches. Once the walnuts have fallen to the ground, they should be gathered as soon as possible. A black walnut hull does not split, but these walnuts will drop from the tree as they mature. Once the trees begin to release their fruits, the rest of the harvest can be knocked loose from the tree using the same technique as with the cultivated walnut. Collect the walnuts off the ground immediately to avoid mold or insect invasion. A specialized nut-gathering tool can be used to quickly collect the walnuts from the ground, and a number of different models are available at a wide range of prices.

Post-Harvest Processing

Due to their high tannin levels, the longer walnuts remain in their hulls, the more bitter they will become; therefore, it's important that the nuts have their hulls removed as soon as possible after harvesting. Since English walnuts split open at maturity, these hulls can easily be removed by hand. Black walnut hulls are more difficult to remove. This job can quickly become a tedious task and the yellow juice of the fruits will stain your hands and clothing. This chore can be simplified with the use of a tabletop-mounted corn-shelling machine. The walnuts can be fed into the hand-cranked device just as one would use it for shelling ears of corn, making short work of walnut hull removal. Hulled walnuts need to be well washed, using a power washer or high-pressure hose, to remove any remaining bits of hull and fruit residue. They can also be agitated in a bucket of water, but this technique will take longer as fewer walnuts can be washed at one time. Once the nuts are clean, they should be laid out to dry. It's best if walnuts can be left in a cool, dark, well-ventilated area to cure for at least two to three weeks. Test their readiness by cracking open a nut to test how firm the kernel inside is. A soft and rubbery walnut has not finished curing, but if the nut is firm and snaps, your walnuts are ready.

Walnuts will need to be shelled before being run through the oil press. The technique will vary depending on the species of walnut you are working with. English walnuts have a thinner shell and larger kernels than their wild relatives and can easily be cracked by hand. Various nutcrackers are available on the market including lever-action models that are ideal for breaking the hard shells of black walnuts. Black walnuts, and other hard-shelled species, can be soaked overnight in water, softening the shell to make cracking easier. Separating the broken fragments of shell from the nut meat can be a challenging and time-consuming task. On a small scale this can be accomplished by hand, especially with English walnuts due to their thin shells and large, easily removable seed. Walnuts that are more difficult to crack are likely to result in a mix of shattered shells and small, broken pieces of nut. On a commercial scale, these nuts are cleaned using a color separator, a machine designed to sort and separate seeds and other small items based on their size and color, or with an aspirator, a machine that uses air pressure to separate the lighter shells from the heavier nuts. Building plans are available online for home-use aspirators, and one of these devices could surely be configured to clean walnut seeds.

Pressing

Walnuts can be stored in-shell and will keep for up to three years if kept in a cool, dry location. Once cracked, the shelf life of the nuts decreases dramatically, and for best preservation, they should be stored in a refrigerator or freezer. Refrigerated walnuts will keep for up to three months, while frozen nuts can be stored for a year. Walnuts stored in the freezer should be allowed to warm up to room temperature before being run through the oil press.

Walnuts can be run through the machine raw and will benefit from being crushed before the extraction process, as this larger surface area

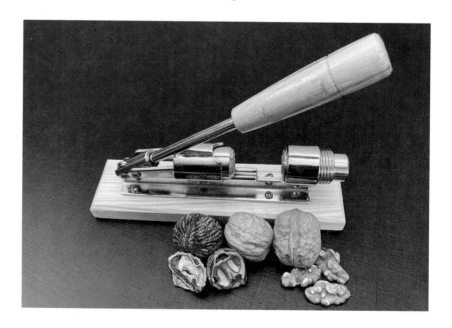

Black walnuts (*left*) and English Walnuts (*right*) with a lever-action, tabletop nutcracker.

will increase yields. The turbid oil should be refrigerated immediately while it settles, and then decanted or filtered to remove any sediment. The nuts could also be roasted before pressing, thus altering the flavor while notably increasing output. Simply bake the nuts at a temperature around 248°F (120°C) for approximately 30 minutes and then run them through the oil press while hot. This is the traditional method for extracting walnut oil in Switzerland.

The seedcake remaining after oil extraction can certainly be used as cattle feed, but this expensive and labor-intensive product is far better suited to culinary applications. Walnut seedcake can be enjoyed in various recipes and is ideal as a supplemental ingredient in cookies and cakes.

Storage

An unopened bottle of walnut oil can be stored in the pantry and will keep for up to a year before rancidity is a concern. Once opened, the oil should be refrigerated, and at temperatures around 45°F (7°C) it will remain shelf stable for around six months. Like other oils, walnut oil should be stored away from direct light in a cool, dark location.

Uses

Cold-pressed walnut oil has a golden amber hue and a light nutty taste, although roasting the seeds before pressing results in a slightly darker-colored oil and deeper flavor. With a smoke point around 320°F (160°C) walnut oil can be used in cooking, although exposure to these temperatures will cause the oil to lose its delicate nutty flavor and take on a bitterness that many find unpleasant. Walnut oil is best suited for cold-culinary applications such as salad dressings and as a finishing oil for various dishes, including steamed vegetables.

Walnut oil is purported to be beneficial in topical applications as a moisturizer to heal dry and cracked skin, although due to its high price, is rarely used as an ingredient in commercial skin care products. Home crafters should certainly consider using walnut oil in their preparations for these purposes. The oil is also useful for sealing wood and is used by many on wooden spoons and bowls as a treatment to maintain their quality.

Unlike the previously discussed oils, which are extracted from the seeds and nuts of various crops, the following three oils are pressed from the flesh of the fruits. Although each of these oilseed crops is certainly different from the others, this particular similarity warrants grouping them together here. In certain climates and locales each of these species would be a viable crop, yet these brief entries do not include growing and harvesting techniques but instead focus on post-harvest processing, pressing, storage and use of the oils.

Avocado   *(Persea americana)*

The avocado is native to south-central Mexico and South America. Although archeological evidence suggests that the fruits were consumed as early as 10,000 years ago, it's widely believed that the domestication and cultivation of the avocado didn't take place until 5000 BCE. The fruits have been utilized as a food source, for religious purposes and as inspiration for pottery and artwork. While Mexico remains the world's largest producer of avocados, California and Peru both boast significant avocado production.

Although it's certainly plausible that avocado oil was enjoyed long ago, the first written documentation of the oil was from the British Imperial Institute in 1918, and in 1934 the California Chamber of Commerce made mention of avocado oil being produced from bruised or otherwise blemished fruits. At the time, the method involved first drying the flesh of the fruit before pressing it in a hydraulic press to extract the oil. This technique is similar to how a small-scale oil producer might press avocado oil today.

In the early 1940s, solvent extraction of avocado oil became common, with the refined product being utilized for machine lubricants as well as cosmetics. Scientists in New Zealand developed a low-temperature, solvent-free process for extracting the oil in the 1990s. This process involves first peeling and pitting the fruits, before mashing and mixing the flesh in a process known as malaxation. This helps the avocado flesh release its oils, which are then completely separated from the mixture by use of a centrifuge. Solvent-extracted avocado oils continue to be used for cosmetics and industrial applications, while the oil extracted via centrifuge is considered superior for culinary purposes.

Pressing

Small-scale production of avocado oil is likely to rely on the more traditional technique of compressing the dried flesh of the fruit. The fruit should be pitted and can be peeled, although this second step isn't required for successful extraction. Drying the fruit should be expedited through use of a food dehydrator, and slicing the avocado into thin pieces is recommended. Once the avocado flesh is well dried, it should be broken down into smaller pieces for ease of pressing. The ideal moisture content for expeller pressing of avocado flesh is around 10%, and this can be achieved by taking the dehydrated pieces of fruit and combining them with a small amount of water. The ideal ratio is one cup (250 ml) of water for every 5⅓ ounces (150 g) of dried avocado. Thoroughly mix the water and avocado, then allow the mixture to rest in a sealed plastic sandwich bag for 24 hours. At this point, the avocado is ready to be pressed. An expeller-style machine is ideal for this job, with the adjustment bolt in place to provide the required pressure.

Storage

Unrefined, expeller-pressed avocado oil has a notably shorter shelf life than oils extracted via centrifuge, but if kept in an airtight container, and refrigerated, the oil should keep for four to six months before rancidity becomes a concern. Allowing the oil to settle, then decanting and filtering the oil to remove any particles will improve storability as well as the smoke point of the final product.

Uses

Unrefined avocado oil has a smoke point of 482°F (250°C), higher than any other unrefined, plant-based oil, making it ideal for high-temperature cooking such as frying, searing and sautéing. The flavor of the oil also lends itself well to cold-culinary applications including marinades and salad dressings and as a finishing oil for roasted vegetables, meats and other dishes. Avocado oil is touted as a healthy cooking oil and is rich in oleic acid and lutein, an antioxidant thought to be beneficial for eye health.

Avocado oil is also included in numerous cosmetics, especially those designed to nourish and moisturize the skin. The high oleic acid content of avocado oil is known to promote collagen production, helping to heal the skin, and is useful to protect the skin from the UV damage associated with sunburn. The oil is also considered beneficial in the treatment of acne and is included in formulations to relieve the inflammation associated with eczema and psoriasis.

Coconut   *(Cocos nucifera)*

Coconut is a member of the palm family, *Arecaceae*, and was first domesticated on the islands of southeast Asia. The inner flesh of the mature seed is the portion of the plant collected for food as well as processed for oil. Although the traditional methods of extracting the flavorful oil from the seeds of the coconut palm are reasonably effective and still in use today, most, if not all, commercially available coconut oil is extracted utilizing modern techniques. There are various types of coconut oil available, each manufactured in a different way. Most of these methods extract the oil from the coconut copra, which is made from the dried inner flesh of the seed.

Perhaps the most common product is simply refined coconut oil. This oil is extracted from the copra and dried kernel using high heat and a hydraulic press. The oil is further processed through bleaching and deodorization. The next most common is hydrogenated coconut oil, which is made by taking the refined oil and combining hydrogen with the unsaturated fats in a catalytic process, thus increasing the melting point of the oil to around 100°F (38°C).

Another version of coconut oil available on the market is fractionated coconut oil, which is made by removing the long-chain fatty acids from the oil, resulting in a product that remains liquid at room temperature. This type of oil has an increased shelf life and is preferred by many cosmetics manufacturers as it is easily blended with other oils and can also be used as a carrier for essential oil blends.

Much of the available coconut oil on the market is solvent extracted through the use of hexane, although expeller-pressed oils, often referred to as virgin coconut oil, are available. To meet the increasing demand for solvent-free plant oil products, the use of a centrifuge to extract coconut oil from a mash of coconut meat and water, a technique similar to that used for avocado oil, is also becoming more common.

Pressing

Expeller-pressed coconut oil is a possibility on a small scale for processors with access to copra, although the yields will be less than what can be achieved through other extraction methods. The copra should be dried to a moisture content of around 10% and then fed into the turnscrew-style expeller press. Heating the copra and the press during production will increase yields. The oil should be well filtered to improve both its quality and shelf life. The resulting seedcake is not ideal for consumption but can certainly be used as a supplemental feed for livestock.

Storage

Coconut oil is quite shelf stable and will keep for up to two years if properly stored. As with other seed and nut oils, coconut oil should be stored in airtight, food-grade containers and kept in a cool, dark location. A cupboard or pantry is ideal if the oil can be kept cool. Unrefined virgin coconut oil will remain solid at temperatures at or below 76°F (24°C). Coconut oil in its solid state is considered more stable than when in its liquid form and can be stored in the refrigerator in warm climates to prolong shelf life. Refrigerated coconut oil will be solid but will liquefy once removed from refrigeration and warmed.

Uses

Virgin coconut oil has a smoke point of around 350°F (175°C), making it a perfect option for baking, and the oil is often used as a substitute for olive oil in sautéing and other similar applications. Coconut oil is sometimes suggested as an alternative to butter and other oils, although its high saturated fat content leads some health experts to warn against frequent use of the product in the kitchen.

Despite the fact that in recent years it has become more popular as a culinary oil, most coconut oil is destined for use in the manufacturing of cosmetics, including refined, fractionated and virgin types. Coconut oil is certainly one of the most widely used ingredients in small-scale and home-based health and beauty products due to its emollient properties, which make it ideal for lotions, creams and other topical ointments. Coconut oil is also used in a number of hair-conditioning products, strengthening the hair, adding shine and relieving dry, itchy scalp and dandruff.

Olive   *(Olea europaea)*

There is debate among scholars regarding the exact origins of olive domestication, although it is agreed that wild olives are native to the Mediterranean region and early cultivation of the crop likely took place some 7,000 years ago. While the small trees are sometimes grown for their wood, the vast majority of olive trees are cultivated for the production of edible fruits, with 90% of these olives being harvested for oil production. Archeological evidence can trace some of the earliest known olive oil production to an eastern Mediterranean area just outside the modern Syrian city of Aleppo.

The traditional method for extracting olive oil involves first crushing and grinding the fruits under a millstone to create a paste. After allowing the paste to rest, it is placed into a cage-style press, similar to the equipment used to press apples or grapes for juice. The resulting product is a blend of oil and vegetal water, which is allowed to settle. The water, which is the denser of the two liquids and therefore the heaviest, will sink to the bottom. The oil is then decanted off the top. Modern olive oil producers make use of a centrifuge to separate the water from the oil, but the traditional method is similar to the technique utilized by the small-scale home producer.

Commercial olive oil is graded and labeled based on quality as well as method of production. Europe, North America and Australia use differing standards and rules to determine how each olive oil product is required to be labeled, but the highest-quality products are consistently labeled "extra virgin." Oils with this label are mechanically pressed, meet the highest standards for flavor and odor, and their acidity levels do not exceed the standard set by the country of production. Mechanically pressed olive oils with minor sensory defects, such as discoloration or odor, and slightly higher acidity levels are labeled "virgin." Lower-quality oils are sometime referred to as "lampante," or lamp oil, and generally require further refinement as they are not meant for human consumption but are instead utilized for industrial applications.

Pressing

Similar to avocado, olive oil is extracted from the fruit of the plant as opposed to the seed or nut. Olives can be pressed for their oil while either fresh or dried. Fresh olives should first be crushed or ground into paste using a food processor or similar equipment. The olive paste should then be gently warmed and mixed for up to an hour in a process known as malaxation to help draw the oil from the fruits. It's critical that the temperature of the mix is kept at or below 80°F (27°C), to ensure the highest quality oil. After malaxation, the olive paste is loaded into a cage-style press while still warm and slowly squeezed to release the oil. This turbid oil will need to settle because it contains the vegetal

Majestic olive trees can produce fruit for hundreds of years.

liquid from the fruits and this water will negatively impact the quality and shelf life of the olive oil. Once settled, the water will have collected at the bottom of the container, and the oil can be decanted from the top, filtered and bottled.

Oil can be extracted from dried olives using a turnscrew-style expeller press. Fresh olives, including the pits, should be crushed and then dried using a food dehydrator. Once completely dry, the olives should be combined with water and stored inside a plastic bag for 24 hours to reach the proper moisture level for pressing. Use 3½ ounces (100 ml) of water for every two pounds (1 kg) of dried olives to achieve 10% moisture before extraction. At this point, the crushed olives can be run through the press, utilizing the adjustment bolt to provide the necessary pressure. While olive oil produced in this fashion will not contain the vegetal liquid, it will still need to rest for 24 to 48 hours after extraction for the small particles of fruit to settle before being decanted.

Storage

Similar to most unrefined oils, olive oil should be stored in a cool, dark location away from direct light. The optimal temperature for keeping your oil fresh is 50° to 60°F (10°–15°C), so a cool pantry or cupboard is ideal. Although it is common for kitchen oils to be stored near the stove, this area of the kitchen is the warmest, and storing oils here should be avoided in order to maintain their longest potential shelf life. Exposure to air will also expedite rancidity, so consider filling a small bottle for use in the kitchen while storing a majority of your oil supply in a separate location, ideally in a dark area away from any potential heat source. Once opened, unrefined virgin olive oil should be consumed within six to eight months. If unopened, and properly stored, olive oil will keep for up to two years before rancidity becomes a concern.

Uses

Olive oil is perhaps the most common culinary oil. The smoke point of unrefined, virgin olive oil is between 375° and 400°F (190°–205°C) and the oil is frequently used for sautéing, stir frying, and other high-heat applications, although the delicate flavors and nutritional qualities of unrefined olive oil can be negatively affected by such high temperatures. Cold-culinary uses for olive oil include as a salad dressing ingredient, marinades, finishing and dipping oils or to flavor raw seafood dishes.

Olive oil is also widely used in topical products such as lotions, face creams, wrinkle treatments and salves to ease the discomfort of a sunburn. The oil exhibits antibacterial properties and can be included in formulations for this purpose as well. There are numerous hair conditioning products available on the market that employ olive oil for its moisturizing qualities.

Additional Seeds and Nuts for Consideration

Many different seeds and nuts can be pressed for their oil.
Pictured here (*left to right*): pinenut, corn, cashew, squash seed, pistachio.

Amaranth (*Amaranthus* spp.)

There are approximately 75 species of amaranth, yet only two of them are commonly used for seed oil production, *A. cruentus* and *A. hypochondriacus*, both of which are commonly grown as grain crops. Amaranth seeds consist of 6%–9% oil, which is low for an oilseed crop but notably higher than most cereal grains. A majority of commercial amaranth seed oil is solvent extracted, typically through the use of hexane, but cold-pressed options are available. Amaranth oil is prized for its high squalene content, a compound used in many formulas to hydrate skin and hair.

Apricot (*Prunus armeniaca*)

Apricot kernel oil is extracted from the inner seed of this edible tree fruit. The kernel contains 40%–50% oil and is commonly expeller pressed. Grinding the kernel before extraction will help the seed pass through the machine and increase yields. The oil is similar to that of almond and peach kernels in both appearance and flavor. The seedcake is utilized separately to extract essential oil via the distillation process. Apricot kernel oil is often used in skin care products and massage oil and as a carrier for essential oil blends. Apricot kernel oil is not recommended for internal use.

Argan (*Argania spinosa*)

Argan oil is extracted from the nut of a tree endemic to Morocco. The oil content of the seed varies anywhere from 30%–50%. Extracting the seeds from the fruit can be an arduous process that involves first drying the fruit and removing its flesh before curing and cracking the nuts to release the kernel. Traditionally, goats would be allowed to eat the fruits, and the nuts would later be gathered from their droppings, thus allowing the animal to handle a large portion of the processing. The argan kernels are then crushed and roasted before being put through the oil press. The resulting oil is used topically in numerous commercial skin care products but is also edible and can be enjoyed in many cold culinary applications, although it is not typically used for cooking as the oil will easily burn.

Beechnut   (*Fagus sylvatica*)

Beechnut oil is extracted from the seed of the European beech tree, which is widely cultivated in North America as an ornamental. Beechnuts contain anywhere from 35%–50% oil. When mature, the nuts will drop from the tree, encased in a spiny husk. The husk will split as it dries, making extraction of the nut an easy task. Shelling beechnuts can be time consuming, but roasting the nuts first will help loosen the shell, making them easier to remove while also improving the flavor of the oil. Beechnuts can be run through a motor-operated press while still in the shell, and the resulting presscake can simply be discarded or offered to livestock. The oil pressed from beechnuts is highly regarded among high-end chefs and is frequently used in baked goods and stir fries.

Black Currant   (*Ribes nigrum*)

A majority of black currant seed oil available on the market is pressed from seeds retained from the production of juice, jelly and jam products. A modern oil-extraction technique utilized to maximize yields from berry seeds while avoiding the use of potentially toxic chemicals such as hexane is known as supercritical fluid extraction. For this process, the seed is ground into a powder and exposed to high-temperature, high-pressure CO_2, thus breaking down the plant material and facilitating the extraction of the oil. Black currant seeds contain anywhere from 27%–33% oil and can certainly be cold pressed, if a reasonable quantity of seeds can be acquired, resulting in an oil commonly used as a nutritional supplement.

Borage 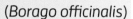 *(Borago officinalis)*

Borage is a common garden herb, grown for its attractive flowers and edible leaves. The oil extracted from its seeds has become a popular nutritional supplement, but care should be taken as regular or excessive use of this oil could potentially result in liver damage. Borage seed oil should not be taken by pregnant women. Most commercial borage seed oil is produced through solvent extraction or superficial fluid extraction, but the seeds contain approximately 38% oil and can be expeller pressed, albeit for notably smaller yields. The oil is used topically for eczema, itchy skin conditions and dermatitis

Brazil Nut *(Bertholletia excelsa)*

Brazil nut trees grow in Peru, Venezuela, Ecuador, Colombia and, of course, Brazil and can reach heights up to 165 feet (50 m). The nuts are approximately 65% oil and can be expeller pressed, resulting in good yields. Due to the size of the nuts, they should be broken or crushed before extraction. If importing brazil nuts for oil extraction, consider purchasing already shelled, broken nuts as the shelling process can be quite tedious. Brazil nut oil is renowned as a hair treatment, adding shine and body, while also treating dandruff and other scalp conditions. The light-bodied oil is also included in numerous skin care products. Although the oil is edible it should be consumed only in moderation due to the nut's high selenium content, which can be toxic in high doses.

Butternut Squash *(Cucurbita moschata)*

Much like pumpkin seeds, butternut squash seeds can be pressed for their flavorful and nutritious oil. Butternut is simply a variety of *C. moschata*, and any seeds from this species produce quality oil. In fact, the seeds of most squash species are edible and can be used as oilseed crops, although the flavor profiles and yields will vary from one variety to the next. Squash seeds contain anywhere from 10%–30% oil, with butternut being one of the best-yielding cultivars. Roasting the seeds before pressing will notably increase yields and add robust flavor to the oil. Butternut squash seed oil has a smoke point of 425°F (218°C) and can be used for baking, sautéing and grilling as well as low-temperature applications such as salad dressing, marinades and flavorings for yogurts and smoothies.

Carrot   (*Daucus carota*)

There are three significantly different products that are referred to as carrot oil or carrot seed oil, with the most common being carrot seed essential oil, which is steam distilled from the seeds of wild carrot, commonly known as Queen Anne's Lace. The second is made by infusing macerated carrot root in a carrier oil. But the product relevant to the small-scale seed and nut oil producer is made through expeller extraction. This oil is also derived from wild carrot seeds, which contain approximately 8% oil. Carrot seed oil is used in commercial beauty products, especially facial cleansers and toners. The oil is not considered safe for internal consumption.

Cashew   (*Anacardium occidentale*)

The seeds of the cashew tree average around 47% oil and can be mechanically pressed utilizing a turnscrew-style expeller or a cage press similar to the equipment used to press apples or grapes for fermentation. The resulting oil is a deep yellow color and very flavorful, especially if the nuts are roasted before pressing. Due to the relatively high cost of cashews, the oil is considered precious and is rarely used in large quantities or for baking and frying. Cashew seed oil is ideal as a salad dressing ingredient or as a finishing oil to add flavor to stir fries, roasted vegetables and other dishes. The oil is featured in a bevy of beauty products that boast its usefulness in moisturizing and smoothing skin, and it's also included in a number of hair care formulations where it is said to slow balding and prevent gray hair.

Coriander   (*Coriandrum sativum*)

Not to be confused with the more common coriander essential oil, which is produced through steam distillation, this product is expeller pressed from coriander seeds. These seeds contain 10%–15% oil, and the resulting product is often utilized as a nutritional supplement. Although the oil is quite strongly flavored, some recipes do call for its use in smoothies, guacamole, hummus and salads. The oil is quite sensitive to heat and should be used only for low-temperature applications.

Corn   (*Zea mays*)

One of the most common mass-produced seed oils, corn oil is extracted from the germ of the corn seed. Almost all corn oil available on the market is produced through a combination of first expeller pressing and then solvent extraction. The oil is refined through degumming, bleaching and deodorizing, creating a product with a high smoke point and minimal flavor, ideal for high-heat cooking. Some specialty producers do offer unrefined, expeller-pressed corn oil, though it is notably more expensive due to the much lower yields. The corn germ contains 35%–55% oil, but the germ must be separated from the seed before pressing.

Cottonseed   (*Gossypium* spp.)

Cottonseed oil is a mass-produced product extracted from the seed of various species, including *G. hirsutum*, *G. herbaceum* and *G. barbadense*. Hulling the seeds requires special equipment, and the oil can be extracted via expeller press, although most of the oil available on the market is solvent extracted. Cotton seeds contain 15% and 20% oil. Unrefined cottonseed oil contains the toxic compound gossypol and is not recommended for human consumption, although the oil is often utilized as an insect repellant.

Gourd   (Various species)

A number of different oils are offered under the name gourd seed oil, each pressed from a different species, although they are all often used in similar manners. Each of these oils is reputed to be beneficial for the skin and is included in lotion and balm formulas for this purpose. These oils are also used as carriers for topical essential oil blends. All of the gourd seed oils can be produced via expeller extraction methods. A few of the most common species used in the production of gourd seed oil include *Lagenaria siceraria*, also known as the bottle gourd; *Luffa aegyptiaca*, the sponge gourd; *Cucurbita foetidissima*, the buffalo gourd; and *Momordica charantia*, the bitter melon.

Grapefruit   (*Citrus × paradisi*)

There are three distinct extractions available that are all made from the grapefruit, but only one of these is worth consideration for the small-scale oil producer: expeller-pressed oil from the seed of the fruit. The seed is approximately 30% oil, but yields through methods other than solvent extraction are comparatively low. The other products available are grapefruit essential oil, extracted from the peel of the fruit, a process used for other citrus essential oils as well, and grapefruit seed extract, which is made by infusing the ground seeds in a menstruum, typically either alcohol or glycerin.

Jojoba   (*Simmondsia chinensis*)

Jojoba oil is common in topical beauty and wellness products, in particular lotions and balms designed to moisturize the skin. The plant is native to the southwest United States and northern Mexico, and the seeds contain anywhere from 30%–60% oil, although the average content is around 50%. Jojoba seeds should be well dried, which will take up to two weeks, before running them through the oil press.

Expeller-press extraction will yield approximately 75% of this oil with the first pressing, and a second pressing of the seedcake will result in an additional 5%–10%. Due to jojoba's chemical structure, it will become cloudy, thicken and eventually solidify into a wax at temperatures around 50°F (10°C) or below.

Kenaf   (*Hibiscus cannabinus*)

A plant related to okra and cotton, kenaf is most widely cultivated for its fiber, which is made into everything from cordage, cloth and paper to resin-based plastics and automobile parts. The seeds average an oil content of 20% and can be expeller pressed to extract an edible oil. The oil is used in cosmetics as well as in the production of biodiesel fuel. Unlike the related cottonseed, kenaf oil does not contain the toxic compound gossypol and is therefore considered safe for human consumption. With its high levels of polyunsaturated fats, kenaf seed oil is best suited to cold culinary applications.

Macadamia   (*Macadamia integrifolia*)

Most, if not all, of commercial macadamia nut oil available on the market is pressed from the seeds of *M. integrifolia*, a species of tree native to Australia, but oil of similar quality can be extracted from *M. tetraphylla*, the rough-shelled macadamia native to northeastern Wales. The seeds can contain up to 75% oil, although 65% is more common, and can be cold pressed raw or roasted before being run through a turnscrew expeller press. The oil pressed from raw seeds is typically lighter in flavor and very light yellow in color, while the roasted nuts produce a slightly darker and more earthy oil. Macadamia nut oil has a high smoke point of around 410°F (210°C), making it ideal for cooking, sauté and baking applications. The oil is also employed by the skin care industry, thanks in part to its high vitamin E content.

Mongongo   (*Schinziophyton rautanenii*)

This large, spreading tree is distributed throughout subtropical southern Africa, where the nuts have been used as a food source for centuries. Mongongo trees, also referred to as manketti, produce a velvety, egg-shaped fruit that contains a thick-walled pit within which can be found the highly nutritious nut. The pits are often roasted in order to weaken the shell, making them easier to crack, or sometimes the nuts are simply gathered from elephant droppings, cleaned and then processed. The seeds contain approximately 60% oil and can be crushed and run through an expeller press. The resulting oil is edible but most frequently used topically to soothe severely dry skin, being widely considered beneficial for all skin types. The oil is also thought to protect the skin from UV damage and has been used for this purpose in Africa for centuries. Mongongo oil is rich in vitamin E, and therefore is considered shelf stable, even when stored at room temperature.

Neem   (*Azadirachta indica*)

Neem oil is extracted from the seeds of an evergreen tree native to the Indian subcontinent. The seeds contain an average of 40% oil and can be expeller pressed, although a number of commercial products containing neem oil are made through solvent extraction. The oil has a harsh, unpleasant odor and is commonly used as a pesticide. Neem oil repels a wide variety of pests including aphids, cabbage worms, leafminers, whiteflies, mites and Japanese beetles. Neem oil is potentially toxic and should never be ingested.

Okra   (*Abelmoschus esculentus*)

Okra seeds contain anywhere from 12%–20% oil, depending upon culti-var, though the average content is around 17%. A majority of okra seed oil is solvent extracted and is considered a reliable substitute for cottonseed oil in the manufacturing of shortening and margarine. The seeds can be expeller pressed, with noticeably diminished yields, and the result is a flavorful oil that can be enjoyed in many low-temperature culinary applications. Much like the closely related cottonseed, okra seeds contain the toxic compound gossypol, although at notably lower levels. Numerous studies have explored the potential use of gossypol as a contraceptive for men due to its effect on the male reproductive system. Okra seed oil is also considered ideal for soap making as it is a rich source of unsaturated fatty acids.

Palm   (*Elaeis guineensis*)

There are two distinct oils extracted from palm; the first is pressed from the flesh of the fruit, while the other, known as palm kernel oil, comes from the seed of the tree. Most of the world's palm oil is taken from the African palm, *E. guineensis*, although, to a lesser extent, the American palm, *E. oleifera*, is also used. Mass-produced palm oil involves an elaborate system of milling, refining, fractionation and crystallization, but the traditional methods of extraction are simple enough to be done at home. Simply wash and then boil the fruits for twenty minutes or so to loosen the flesh from the seeds. Using a food processor or similar equipment, agitate the warm fruits to separate the hard nuts. The flesh of the fruit can then be squeezed, either by hand or in a cage-style press, similar to what one would use to press grapes, to extract the oil from the fruit. The kernels can also be removed from within the nuts and run through an expeller press to produce palm kernel oil. Palm oils are some of the most widely used oils in the world and can be found in soaps, cosmetics, processed foods, biofuel, as well as countless industrial applications. The high demand for these oils has caused dramatic negative impact on the trees' native ecosystems as countless acres of rainforests have been burned down, making room to expand palm plantations.

Pine Nut   (*Pinus* spp.)

All species of pine produce an edible nut, but only around 20 of those species develop nuts large enough to be considered valuable for human consumption. Pine nut oil can be pressed from the seeds of any of these species, although the common American and European pine nut oils are most typically used for culinary application, while the Siberian and Korean types are often utilized for medicinal purposes. This is due to the higher levels of pinolenic acid and antioxidants found in the seeds of Siberian pine compared to other species. Pine seeds contain up to 60% oil and are often cold pressed in order to retain the oil's beneficial qualities. The oil has a very low smoke point and is generally used for low-temperature cooking as well as a nutritional supplement occasionally found in health food stores.

Pistachio   (*Pistacia vera*)

The pistachio is native to the Middle East and central Asia and is closely related to the cashew. The oil extracted from pistachio seeds is considered a delicacy by many high-end chefs. Pistachio seed oil has a strong nutty flavor, especially when the nuts are roasted before pressing. The seeds contain 50%–62% oil and can be cold pressed to produce the highest-quality product. Although the oil is quite flavorful it is still delicate and should be used only in cold-culinary applications. The seeds and their oil are also typically expensive to purchase and should be used sparingly as a finishing oil for salads, meats and other dishes.

Pomegranate   (*Punica granatum*)

The pomegranate is an ancient fruit with a long and storied history. Native to Persia and east into the Himalayas of northern India, the fruits have been touted for their nutritional value since ancient times. The oil pressed from the seeds of the fruit is used topically and is thought to be beneficial for the skin, hair and overall health. Pomegranate oil is incorporated into cosmetics and wellness products and is also used as a carrier for essential oil blends. The seeds contain anywhere from 7%–16% oil and are typically cold pressed.

Rice Bran   (*Oryza sativa*)

Rice bran is the byproduct of rice milling, the process that changes brown rice into white. Essentially, the bran is the hard outer husk of the rice seed from which rice bran oil can be extracted. This oil is typically used for frying, sautéing and other high-temperature applications thank to its relatively high smoke point of 450°F (232°C). Although the husk contains 17%–21% oil, traditional expeller pressing of the bran offers lower-than-ideal yields. For this reason, a majority of all commercial rice bran oil is extracted through the use of solvents, typically hexane, although promising results have also been found through supercritical fluid extraction, offering a product considered to be superior.

Soybean   (*Glycine max*)

Although native to East Asia, soybeans are now grown around the world, especially in the United States, as a commodity crop. The oil extracted from soybean seeds is one of the most widely consumed cooking oils, and the seedcake resulting from the extraction process is used as a high-protein feed supplement for cattle and other livestock. Soybeans contain an average of 18%–20% oil, which is typically extracted through the use of solvents such as hexane. Soybeans can be expeller pressed, although the yields are remarkably lower. This can be somewhat improved by heating the seeds to temperatures of 140°–190°F (60°–88°C) before pressing. Cold-pressed soybean oils are available, although for a notably higher price due to the lower yields this method produces. Although cold-pressed soybean oils retain their natural flavor and nutritional content, they have a significantly lower smoke point than the solvent-extracted oils and therefore should be used only for cold-culinary, as well as cosmetic, applications. Oils extracted from the wild soybean, *G. soja*, are also sometimes available and are typically employed for massage or other topical uses.

Watermelon · (*Citrullus lanatus*)

Native to Africa, watermelons were likely domesticated more than 4,000 years ago. The oil extracted from the seeds of these fruits has been used topically in cosmetics and as an emollient but most notably to protect the skin from sunburn and UV damage. In Africa, watermelon seed oil is commonly known as Ootanga oil and occasionally as Kalahari oil. Some solvent-extracted watermelon seed oils are available on the market, but a majority of products include expeller-pressed oils. The seeds' average oil content is around 25%, and higher yields can be expected from seeds that have been roasted before pressing, although cold-pressed watermelon seed oil is considered superior for use in beauty and wellness products. The seeds can be shelled before extraction but can easily be pressed in-shell, especially with a motorized or pedal-powered machine.

In Conclusion:
Beyond the Press

THERE MAY COME A TIME when you feel ready to move beyond small-scale production for home use and perhaps venture into the commercial realm of oil extraction. This could be after many years of experience evaluating various oilseed crops and equipment, or commercial production may be the original motivation that inspired you to begin this journey in the first place. Regardless, commercial production will certainly require additional consideration above and beyond what is needed for personal or family oil pressing. Again, you need an honest evaluation of your needs, goals and resources to construct a successful plan for expanding your operation. Many factors must be considered, but for commercial production, perhaps the place to begin is an investigation into the licensing and permit requirements for production of food products in your area. Scaling up to a commercial level may also require a dedicated space or facility for pressing, as well as for storing seeds, nuts and, of course, the finished oils. This new space will need to house bottles and bottling equipment and will likely need to be much larger than the area needed for smaller-scale home production setups. This expansion may also necessitate upgrades in pressing, harvesting and processing equipment.

In addition to this, you will need to develop a plan regarding your oilseed supply. Will you be growing and harvesting your own seed crops or sourcing from a local grower? Understanding the size and quality of your seed supply will help you to determine your product line. Will you be able to offer certain oils year round, or will your offerings be strictly seasonal, based on the availability of the oilseeds? Develop a business plan to help you work out these details well in advance of taking the leap into a business startup. Of course, funding the expansion will also require capital for equipment, seeds, licenses, labeling, marketing, etc. A well-thought-out business plan may help to secure this funding from a bank or other financial institution.

Owning and operating a commercial oil business, no matter how small, is a significant departure from simply working to provide for a household's needs. It will demand far more of your time and energy, yet is not guaranteed to be a financial success despite this effort. While passion and hard work certainly play a role in a commercial venture's success, they are not the only factors that determine a favorable outcome. Risk comes with all ventures, and analyzing this risk to determine its worth is critical. Running a business is not for everyone, and while some may be content hitching their fortune to the will of consumers, others might find that focusing their attention on personal production may be the most rewarding route.

An interesting model worth considering for a startup commercial oil production venture is that of the cooperative, or co-op. Cooperative businesses, although perhaps not as common as the more traditional business structure, have existed in some form or another since at least 1844, when a group of 28 artisans who worked in the cotton mills of Rochdale, England, decided to pool their resources in order to acquire basic goods at a substantially lower price. By combining their resources, they were able to achieve greater buying power, essentially allowing them to purchase basic goods such as flour, sugar and eggs at wholesale prices. Over time they were able to expand their membership, thus increasing their available capital as well as their buying power. In essence, this member-owned business, The Rochdale Society of Equitable Pioneers, became the first cooperative grocery store and the model for future cooperative endeavors.

Today, cooperatives operate in a wide range of industries, from farming and art to childcare, housing and student organizations, but they all tend to follow similar ideals and principals: member owned and operated, voluntary membership, autonomy and independence, equitable and democratic economic participation and concern for the community. The agricultural cooperative is worth examining further as a potential model for a community cooperative built around seed and nut oil extraction. Often these co-ops offer specific functions to their membership, be it marketing, supplies or services, although there tends to be much overlap between these fields and a cooperative based on oilseed pressing would likely best succeed with attention paid to all three. A marketing cooperative assembles, packages and sells the products crafted by the membership, while a supply co-op could focus on pooling member resources to acquire quantities of seeds, nuts, bottles and other items necessary for an oil pressing company to operate. A service-based cooperative would specialize in harvesting, processing or hulling various seeds and nuts in preparation for extraction, which may not be feasible or economical for the individual to handle on their own.

A community-led, cooperative oil pressing operation could be integrated into a garden program, local food initiative, intentional community, fruit and nut growers' association or other similar organizations. The oil pressing could be used as a way to generate revenue for the organization and its programming or as a way to simply provide the members with high-quality, locally produced seed and nut oils. Additionally, these cooperative endeavors could be utilized as community outreach and education opportunities, to possibly facilitate membership expansion while introducing more people to a production-centric DIY lifestyle. The startup considerations for a cooperative model would be similar to an independently owned and operated business, but the collaboration and consolidation of resources among co-op members would certainly ease much of the initial burden.

Reviving traditional artisan skills such as small-scale seed and nut oil pressing revitalizes our relationship with our food system. When locally grown oilseed crops are at the foundation of the operation, in a way this restores and strengthens our sense of place. Whether we work as individuals on our homesteads or as a cooperative through our community gardens, this hands-on, small-batch process connects us to our land and to each other in ways far beyond what we can hope to achieve merely as consumers. Just as the seasonal flavors found at our local farmers markets are superior to any imported or mass-produced commodity produce and the distinct terroir of the finest grapes is reflected in each vintage of wine, so too are the flavors and distinct qualities of oils pressed from locally harvested or foraged seeds and nuts: unrefined, unbleached, in their most natural and perfect state. As we head out into our forests, fields and orchards, let us remember the countless generations of dedicated producers that have come before us, and as we gather our oilseed crops and begin to turn the screws on another batch of precious oil, let us set our hands and our minds to the critical task of preserving these skills and these resources for the next generation of producers to follow in our footsteps.

Index

About the Author

BEVIN COHEN is an author, herbalist, gardener, seed saver and educator. He is owner of Small House Farm, a sustainable herb farm, and he offers workshops and lectures nationwide on the benefits of living closer to the land through seeds, herbs, and locally grown food. He has published numerous works on these topics, including *From Our Seeds & Their Keepers*, *Saving Our Seeds*, *Salvando Nuestras Semillas*, and *The Artisan Herbalist*. He is the founder and president of the Michigan Seed Library Network and serves on the advisory council for the Community Seed Network, a multinational education and networking platform. He lives in Sanford, Michigan.

ABOUT NEW SOCIETY PUBLISHERS

New Society Publishers is an activist, solutions-oriented publisher focused on publishing books to build a more just and sustainable future. Our books offer tips, tools, and insights from leading experts in a wide range of areas.

We're proud to hold to the highest environmental and social standards of any publisher in North America. When you buy New Society books, you are part of the solution!

At New Society Publishers, we care deeply about *what* we publish—but also about *how* we do business.

- All our books are printed on 100% **post-consumer recycled paper**, processed chlorine-free, with low-VOC vegetable-based inks (since 2002). We print all our books in North America (never overseas)

- Our corporate structure is an innovative employee shareholder agreement, so we're one-third employee-owned (since 2015)

- We've created a Statement of Ethics (2021). The intent of this Statement is to act as a framework to guide our actions and facilitate feedback for continuous improvement of our work

- We're carbon-neutral (since 2006)

- We're certified as a B Corporation (since 2016)

- We're Signatories to the UN's Sustainable Development Goals (SDG) Publishers Compact (2020–2030, the Decade of Action)

To download our full catalog, sign up for our quarterly newsletter, and to learn more about New Society Publishers, please visit newsociety.com

ENVIRONMENTAL BENEFITS STATEMENT

New Society Publishers saved the following resources by printing the pages of this book on chlorine free paper made with 100% post-consumer waste.

TREES	WATER	ENERGY	SOLID WASTE	GREENHOUSE GASES
36	**2,900**	**15**	**120**	**15,600**
FULLY GROWN	GALLONS	MILLION BTUs	POUNDS	POUNDS

Environmental impact estimates were made using the Environmental Paper Network Paper Calculator 4.0. For more information visit www.papercalculator.org

Certified **(B)** Corporation

new society
PUBLISHERS
www.newsociety.com

MIX
Paper from responsible sources
FSC www.fsc.org   FSC® C016245

SDG PUBLISHERS COMPACT